IEE MONOGRAPH SERIES 4

PER-UNIT SYSTEMS: WITH SPECIAL REFERENCE TO ELECTRICAL MACHINES

M. R. HARRIS   P. J. LAWRENSON   J. M. STEPHENSON

PER-UNIT SYSTEMS
WITH SPECIAL REFERENCE TO ELECTRICAL MACHINES

CAMBRIDGE AT THE UNIVERSITY PRESS 1970
Published in association with
THE INSTITUTION OF ELECTRICAL ENGINEERS

Published by the Syndics of the Cambridge University Press
Bentley House, 200 Euston Road, London N.W.1
American Branch: 32 East 57th Street, New York, N.Y.10022

© The Institution of Electrical Engineers 1970

Library of Congress Catalogue Card Number: 70-118857

Standard Book Number: 521 07857 1

Printed in Great Britain
at the University Printing House, Cambridge
(Brooke Crutchley, University Printer)

CONTENTS

Preface *page* vii

Foreword xi

1 Introduction 1
   1.1 List of symbols 1
   1.2 General 3
   1.3 Advantages of a per-unit system 7
   1.4 Review of literature 9

2 Basic analysis of coupled circuits 16
   2.1 Single-phase transformer 16
   2.2 Generalisation to multiple circuits 22
   2.3 Turns ratio and leakage reactance 27

3 Circuit analysis of machines 36
   3.1 Transformations and principles 36
   3.2 Secondary-current bases of the synchronous machine 63
   3.3 Machines other than synchronous 77

4 Mechanical and other considerations 90
   4.1 Mean and instantaneous power: r.m.s. and peak bases 90
   4.2 Concept of associated per-unit parameters 92
   4.3 Flux linkage, instantaneous torque and per-unit time 94
   4.4 Inertial torque 97

5 Recommendations and conclusions 99

6	Appendixes	*page* 103
	6.1 Transformation of impedance matrix from 3-phase to 2-axis parameters	103
	6.2 Ideal ratio and leakage reactance	105
	6.3 Equivalent circuits of synchronous machines with nested damper windings	108
7	References	113
Index		115

PREFACE

The importance of per-unit systems in the field of electrical-machine and power-system studies is well known, and, in future, as the emphasis on system and transient problems continues to grow, this importance will increase.

There have been many authors who have used or discussed per-unit systems, but their approaches have differed significantly and there is no literature which satisfactorily discusses or correlates these differences. Additionally, in a number of cases, statements or derivations have been made which are at least misleading and sometimes incorrect. The situation is further aggravated by the fact that most discussions reflect the too-ready assumption that per-unit systems are self-evident, requiring a minimum of thought and explanation. As a consequence, the whole subject appears diffuse and confused, and it has been impossible to exploit the full benefits that the universal use of a clearly defined and standardised system could bring.

This Monograph is designed, in the light of the above situation, with two broad objectives in mind: first, to provide a critical review of the subject and to clarify the similarities of, and differences between, earlier treatments, and secondly, to provide in a single, unified treatment a wide-ranging discussion of per-unit systems, including deeper discussion of several familiar topics and the introduction of new ones, where necessary, to yield a sound and coherent presentation.

The discussion is everywhere rigorous, and mathematical formulation is employed as necessary, but particular emphasis is laid on the physical implications of the mathematics and the systems discussed. The ability to illuminate the physical nature of a device or system is an important part of the value of a good per-unit system and, largely on this basis, we recommend, for synchronous and induction machines, one particular system—the 'full-pitched X_{ad} base' system—which we hope, after proper discussion by the professional institutions, will be universally adopted.

The first chapter of the Monograph introduces the subject, outlining the advantages of per-unit systems, and provides a review of existing literature. In Chapter 2 the basic circuit theory is first introduced very simply through the analysis of the single-phase transformer, and is then generalised in a revealing approach using matrices. The problem of choosing secondary-current bases is clearly interrelated with the basic indeterminacy of the turns ratio and leakage reactance. This leads to the introduction of the concepts of the 'ideal' field pattern and turns ratio, which put into perspective the usual approach in design and analysis, and emphasise the need for care in the general application of this approach.

Chapter 3 deals with the more general circuit problems occurring in rotating machines. These include a coherent derivation of the per-unit, 2-axis equations, a full discussion of the synchronous machine with the various possible bases (particularly for rotor-circuit currents) and the principles on which these depend, and brief treatments of the application of per-unit systems to induction and commutator machines, largely neglected topics.

Chapter 4 discusses the mechanical equations, questions concerned with flux linkage, instantaneous torque and the normalisation of time, and also the choice between peak and r.m.s. quantities. Overall discussion and recommendations appear in Chapter 5, and the Appendixes are devoted to the transformation of the impedance matrix from 3-phase to 2-axis parameters, the ideal turns ratio and leakage reactance, and complete equivalent circuits for synchronous machines with nested damper circuits.

From the above the reader will realise that for a full study of the Monograph, some knowledge of the unified theory of machines is necessary, but much of the argument is accessible given only a familiarity with elementary circuit theory.

In preparing this Monograph we have been much helped by the co-operation of a number of people, and we would particularly wish to thank the following:

D. Ashworth and P. J. Markey of the then Associated Electrical

Industries Ltd, G. Campbell of C. A. Parsons & Co. Ltd, A. J. T. Timberlake and A. Williams of the then English Electric Co. Ltd and K. C. Parton of the then General Electric Co. Ltd.

We are also grateful to Mrs P. Jeffery and Miss L. Myers for the careful and extensive typing of the several versions of the manuscript.

M. R. HARRIS
P. J. LAWRENSON
J. M. STEPHENSON

FOREWORD

A mathematician's problem may be solved to his satisfaction if he can find a relationship between a set of symbols. A physicist may be content with some kind of conceptual 'model'. The engineer learns from the mathematician and appreciates the concepts of the physicist, but he has to make real things—such as electrical machines that actually work.

The theories that the engineer invents to 'explain' the things he makes are always far simpler than the things themselves. An electrical machine is a highly complex electromagnetic device, operating on energy-conversion principles that embrace every detail, down to the atoms that make up the iron, copper and insulant, and the lubrication and cooling materials. The engineer must simplify. Electron spin and molecular friction become mass effects of magnetisation and loss, complicated field patterns are reduced to circuit 'parameters', electromagnetic effects are represented macroscopically by terminal voltages and currents. The engineer has no choice; his models must be greatly simplified, and what he does not know he has to guess.

The guesses do not have to be blind. Exploiting the human ability to press imagination into the abstract, the engineer, in the face of a superabundance of physical causes and effects, may assemble them in lumps to give 'dimensionless parameters'. These, investigated separately, can, within defined limits, make a problem tractable.

One such dimensionless parameter, of an apparently blithe simplicity, is the ratio between similar physical quantities expressed as a percentage or as a fraction; i.e. in per-unit terms. Per-cent value is on a base of 100 and per-unit value is on a base of 1. This sums it up for the uninitiated, who can readily express externals (such as load, voltage, power factor and efficiency) in per-unit terms as easily as in per-cent.

The simplicity is deceptive. When one delves into the internal parts of a machine, awkward questions pose themselves. What *is* 1 p.u. flux linkage or m.m.f. in the face of windings with differing turns, spreads, numbers of phases, voltages and ratings? How are per-unit variables in different reference frames (which may represent the same machine) related? What form is taken by per-unit electromechanical equations?

Interpretations differ, opinions conflict, ambiguities and inconsistencies arise. In this Monograph the authors chart the pitfalls and attempt to make plain the path in a new and comprehensive treatment of a subject certainly in dire need of clarification. It is chastening to find, yet again, that much which is apparently straightforward repays 'another think'.

M. G. SAY

1 INTRODUCTION

1.1 List of symbols

Unnecessary repetition is avoided by noting that voltages, currents and powers shown in lower case represent instantaneous values, and the equivalent upper-case symbols represent time phasors of voltages and currents, and mean values of powers. Any symbol in bold-face type represents a normalised per-unit equivalent. The use of the modulus sign implies the true amplitude (i.e. peak value) of a phasor, and not the r.m.s. value.

Subscripts

a, b, c	three phases
d, q, z	direct-axis, quadrature-axis and zero-sequence
p, n	positive and negative sequence
f, b	forward and backward components (used only in Section 3.3.2)
f	field
k	damper, normally used with an extra subscript to indicate axis on which the damper winding acts, thus kd, kq. (Where n damper windings exist on an axis, k is replaced by digits $1-n$)
l	leakage
m	mutual
o	per-unit base quantity
r	rotor
s	stator
t	transpose of a matrix
α, β	two phases

Main symbols

a	$e^{j\frac{2}{3}\pi}$ forward phase operator
A_{d1}	factor by which maximum flux density is multiplied, to

	obtain the maximum of the fundamental component of flux density with excitation by a sinewave of armature m.m.f. in the direct axis
$[C]$	transformation matrix
D_{d1n}	as A_{d1} but with excitation by the m.m.f. of the nth damper circuit in the direct axis
D_{don}	$\tfrac{1}{2}\pi$ times the factor by which the maximum flux density is multiplied to obtain the average density within the damper circuit of span y_{nd} in the direct axis
F_d	as A_{d1}, but excitation by the m.m.f. of the direct-axis field winding
H	inertia constant
i	current
J	moment of inertia
k	coefficient of coupling
K_d	distribution factor ⎫
K_p	pitch factor ⎬ defined as being $\geqslant 1$
K_ϕ	ratio of the total flux per pole to the total fundamental flux per pole, with excitation by the m.m.f. of the direct-axis field winding at no load
L	inductance, mainly used with double-subscript notation, thus L_{mm} = self inductance of mth circuit, L_{mn} = mutual inductance between mth and nth circuits
$L_{dd}, L_{qq}, L_{ff}, L_{kkd}, L_{kkd}$	self inductances of circuits d, q, f, kd, kq, respectively
L_{df}, L_{dkd}, L_{fkd}	mutual inductances among circuits d, f, kd
L_{qkq}	mutual inductance between circuits q and kq
L_{afd}	mutual inductance between phase winding a and field winding f, when a is centred on the direct axis
L_{akd}	mutual inductance between phase winding a and damper circuit kd, when a is centred on the direct axis
L_{akq}	mutual inductance between phase winding a and damper circuit kq, when a is centred on the quadrature axis
L_d	per-phase direct-axis synchronous inductance defined under conditions of balanced polyphase operation
L_{ad}	armature reaction component of L_d

L_q	per-phase quadrature-axis synchronous inductance
L_{aq}	armature-reaction component of L_q
N	turns ratio, frequently used with single subscript
N_{aa}	series turns per armature phase
N_{ff}	turns per pole for field or full-pitched winding
N_i	ideal turns ratio
p	$\partial/\partial t$
p	power
n	number of pole pairs
R	resistance, used with single-subscript notation, except in problems embodying resistive mutual couplings
s	fractional slip
t	time
T	torque
v	voltage
X	reactance, used with the same subscript notations as L
y_{nd}	span of the nth damper circuit per unit of pole pitch
Z	impedance
δ	torque angle
θ	electrical angle
ω	electrical angular velocity
ψ	flux linkage
ν	mechanical angular velocity

1.2 General

A per-unit system is essentially a system of dimensionless parameters (or groups) occurring in a set of wholly or partially dimensionless equations. Systems of this general kind are used extensively to simplify and illuminate phenomena over a wide range of different physical problems. In some cases, the dimensionless parameters involve relatively complex groupings of dimensional parameters; one widely known example is the Reynolds number in fluid mechanics[1,4]. In other cases, as in the treatment of the circuit aspects of electrical power systems, particularly those of rotating machines—with which this Monograph is concerned—the dimensionless groups mostly have forms which are very simple, nearly all being derived by a process of normalisation in which one physical parameter is divided by another of the same dimensions.†

The quantities involved as denominators in the division process are referred to as 'base' quantities, and they are used because of the ways in which they characterise particular features of the physical system or device. The magnitudes of some may be chosen freely; the magnitudes of others follow by dependence through the laws governing the physical nature of the system. The choice of bases is made so that computational effort is reduced as much as possible, and evaluation and understanding of the main characteristics is made as simple and direct as possible. In practice, this frequently means that the principal per-unit variables assume unit value under rated (full load) conditions.

In the past, many authors have discussed and used per-unit systems in connection with the circuit aspects of machines and power systems. However, as will be explained in subsequent Chapters, their approaches have differed significantly, and there is no literature which satisfactorily discusses or correlates these differences. In addition, in some cases, statements or derivations have been made which are misleading, or even, without proper qualification, incorrect. Also there is generally a too-ready assumption that per-unit systems are theoretically self evident, requiring and receiving only a minimum of explanation; consequently the whole subject appears diffuse and confusing. This Monograph is designed, therefore, partly as a critical review, clarifying similarities and differences between various earlier treatments, partly to present a deeper discussion of a number of matters, and partly to provide, in a single treatment, a wide-ranging discussion of per-unit systems.

It should be noted here that in a straightforward power-system analysis there is an area of application of per-unit systems which is relatively easily understood, and unambiguous. In general, no great problems occur where devices such as transformers and machines are to be represented by simple impedances (referred to their terminals). It is in the detailed analysis of the internal characteristics of devices

† Rather more complex partly dimensionless parameters do arise for machines, although they have not been discussed in quite this context, and would not normally be evoked in the power engineer's mind by the phrase 'per-unit'. One example is the 'goodness factor' discussed by Laithwaite in several of his publications.

(which may be required for transient and unbalanced conditions or in the design process) that there is a greater probability of confusion. This Monograph concentrates mainly on the latter class. Some reference to the former is made, but for fuller treatments the reader is referred to the many available texts[11,14,15,37] on the subject.

The physical implications underlying the mathematics are particularly emphasised in this treatment, since the ability to illuminate these is an important part of the value of a good per-unit system. In so doing, reference will often be made, as in Sections 2.1, 2.3, 3.1.1 and 3.2.1, to the forms of the equivalent circuits which result from the adoption of particular systems. It will be clear, however, that these are of interest because they are diagrammatic representations of an essentially more significant matter; namely the physical nature of the magnetic fields within machines and the manner of their treatment in terms of inductances in the design process.

In the following Sections of this Chapter, there is a short statement of the advantages of per-unit systems, followed by a review of the existing literature. In Chapter 2, the basic circuit theory is simply introduced through the analysis of a single-phase transformer, a generalised and revealing approach to the matter through unified theory is described, and, finally, a discussion of the significance of leakage reactance and turns ratio is presented. In Chapter 3, consideration is given to the more general circuit problems occurring in rotating machines. This includes a coherent derivation of the per-unit, 2-axis equations, a full discussion of the synchronous machine and the various choices of bases, particularly for rotor currents, including the principles on which these depend, and a brief treatment of the application of per-unit systems to induction and commutator machines—a largely neglected topic. Chapter 4 deals with the mechanical equations, flux linkage, instantaneous torque and the normalisation of time, and the choice between peak and r.m.s. base quantities. An overall discussion and recommendations are given in Chapter 5.

Reference has already been made to 'physical' parameters (which are dimensional) and to dimensionless per-unit groups, and in the following Chapters it will be necessary to comment on their nature and the difference between them. The physical parameters which are mainly involved in the discussion are voltage, current, impedance, time, speed and inertia. Since the word 'physical' is useful in other applications, it is not possible to retain it in this context (natural though it is), and an alternative must be found. Rankin[2,3], one of the principal writers on the subject, has suggested 'ampere-inch', but, in addition to its basic unattractiveness, this is now quite unacceptable because of the particular system of implied units. Instead, it is proposed to use the term 'ordinary' hereafter. Thus, taking current as an example, an ordinary current is one measured in amperes. Continuing the example, a current x in per-unit terms, relative to a base current y, is equal to (x amperes/y amperes), which is dimensionless and is properly expressed as x/y p.u. (The nomenclature 'p.u.' will be used only after a numerical quantity; i.e. in place of the units which would occur if the quantity had represented an ordinary parameter. In the text the full term 'per-unit' will be used.)

One other general matter concerning the overall form of per-unit equations should be mentioned here. This form is frequently the same as that of the corresponding ordinary equations, and indeed a per-unit system is often chosen with this feature in mind, so that the per-unit mathematics may have the same general nature as the ordinary mathematics. For example, $v = Zi$ could be an ordinary or a per-unit equation. There is a tendency, however, to suppose that per-unit and ordinary equations will always have precisely the same form, and also that per-unit equations can always be checked for 'dimensional consistency'. In fact neither of these suppositions is universally correct, although in many cases both apply.

It should be noted that the forms of the per-unit power equations are often such as to involve overall coefficients which are different from those in the ordinary equations, and the same may be true of the *equations of transformation* from one reference frame to another. Additionally, a statement such as 'full-load stator i^2R loss = stator resistance' (which is correct in per-unit terms) cannot be 'dimensionally checked'.

1.3 Advantages of a per-unit system

The advantages which can arise from the use of a well designed per-unit system are as follows:

(*a*) The ordinary parameters (current, impedance, losses etc.) of, for example, a particular type of electrical machine, vary widely with variation of the following characteristics: physical size, terminal voltage, power rating, number of phases in a polyphase system, internal connections etc. The per-unit parameters do not, by their nature, depend directly on any of the above,† and are of comparable magnitude for a wide range of machines. A simple inspection of the per-unit parameters immediately reveals much more about the basic nature of a machine than may be observed from the ordinary parameters; for example they reveal efficiency, power factor, stall torque and current as a proportion of their full-load values, relative proportions of core and i^2R loss etc. Thus, a per-unit system embodies a set of dimensionless coefficients which characterises the machine, and simplifies comparisons. This technique is, in general, probably better appreciated and more widely used by mechanical engineers[1,4]. Additionally, certain orders of magnitude become familiar to the designer, and the probability of design errors is greatly reduced.

Not only complete machines, but also individual windings within a single machine, can be meaningfully compared through their per-unit impedances, whereas ordinary impedances reveal little. This is elaborated in Section 3.2.5.

In all comparisons between per-unit quantities, it has to be realised that the same set of base quantities, i.e. the same per-unit system, must be uniformly employed. The need to do this, which may be overlooked, will be clear from the discussion in Chapters 2 and 3. In particular Table 1 indicates the variations in the numerical values of a given parameter when expressed in relation to different base quantities, and hence illustrates the danger of interchanging numerical values between different per-unit systems.

† Strictly, one might say here 'need not depend'. For example, it is possible to have a 2-axis per-unit analysis of the polyphase synchronous machine, in which the field impedance depends upon the number of armature phases (see Section 3.1.1), though such a system is not preferred here.

7

(*b*) The numerical range of per-unit parameters is small, in general being of the order of unity and less. This is valuable for solution by analogue or digital computer, since the variables are of a convenient order. Manual calculations are also simplified.

(*c*) Simplifications occur in the analysis of polyphase circuits under balanced conditions. By defining appropriate per-unit line quantities to correspond with chosen per-unit phase quantities, both line and phase parameters can be represented in one per-unit analysis (and equivalent circuit). This is discussed in Reference 11 and is elaborated in Sections 4.1 and 4.2.

(*d*) In single- and polyphase analysis, the turns ratios of transformers (and the manner of internal connection in the polyphase case) are removed from the analysis, their effects being allowed for by the choice of primary and secondary base magnitudes. A useful consequence of this is that the per-unit self inductance of a circuit is equal to the sum of its per-unit mutual and 'leakage' inductances. (The significance of leakage inductance and its relationship to the turns ratio and the series inductances of the equivalent circuit are, however, less clear cut than this statement might suggest. These matters are fully discussed in Sections 2.3 and 3.1).

(*e*) In the 2-axis theory of the synchronous machine, a per-unit system is useful in removing those arbitrary numerical factors which can appear in the ordinary equations, having values depending on the transformation used. In many treatments, however, a physically incorrect 2-axis transformation is postulated, and the per-unit system then appears to have further advantages in absorbing the difficulties which consequently arise (see Section 3.1.1).

(*f*) A basic set of dimensionless parameters can help to prevent errors in converting performance characteristics between different systems of units. Such conversions may arise in design, or possibly in conforming with units specified by a customer.

(*g*) Specially helpful simplifications occur for variable-frequency conditions, when time is normalised. This normalisation, usual in the American literature, is fully discussed in Section 4.3.

1.4 Review of literature

The following review of the literature on per-unit systems touches on most of the important points, and indicates the different approaches of different authors. It is not intended in any way as a direct criticism of individual authors, but it provides a disconcerting comment on the total picture presented by the existing literature.

The relationship between the stator and rotor per-unit quantities is the first major problem to emerge, and is also one of the most confused. In discussions of this matter, it should be remembered that the choice of a turns ratio between coupled coils is equivalent to the choice of current bases, since the influence of one coil on another depends on the m.m.f. produced; i.e. it depends equally on the turns and the current because of their joint effect. Obviously, in any particular case, the choice can, in principle, be made in an infinite number of ways. Several other substantial difficulties are later evident, particularly with regard to the fundamental 2-axis per-unit transformation, and also various aspects of the treatment of expressions for power and mechanical performance. Only a brief discussion is included at this stage, the various matters being expanded and clarified in later Sections as indicated.

Stephen[5] provides a good introduction to per-unit systems in the application to machines. His contribution is notable for its emphasis of the numerical simplicity of calculations in per-unit terms, and it is fully supported by worked examples. He makes particular reference to the advantages to be gained in the analysis of *induction* machines, and he compares the per-unit parameters of widely differing designs. This application of per-unit systems is scarcely mentioned by other authors, who are generally concerned with the application to synchronous machines.

For the synchronous machine, Stephen selects a unit base of airgap m.m.f. to be that which will excite unit voltage in the armature on open circuit. This might initially appear to be an obvious definition, but it leads to an unfortunate inconsistency; the usually accepted unit armature current (i.e. the rated value) does not establish unit m.m.f. No alternative bases are discussed, but it is stated that a fuller discussion of various topics is necessary.

In two of the most important papers on the subject, Rankin[2,3], restricting himself to the purely electrical aspects of per-unit systems in relation to synchronous machines, summarises the previous work concerned with per-unit impedances, and presents a complete set of design formulas[3]. He points out that two classes of impedance can be distinguished. There are those which are measurable at the machine stator terminals, and those which are not. The former, being independent of the choice of rotor-current base (turns ratio), are unambiguous, but the latter are functions of the current base and therefore become subject to misuse. Rankin goes on to identify the rotor-current bases effectively chosen by pre-1945 authors, and draws attention to the discrepancies between them.

Of the theoretically infinite number of choices of rotor-current bases, he selects four as being of sufficient practical importance for discussion:

(a) *X_{ad} base*

With this base, which Rankin recommends as the best, the base current in any rotor circuit is defined as that which induces in each stator phase a voltage equal to $X_{ad} i_{ao}$. Inherent in this definition, as will be emphasised, is the assumption of a sinusoidal mutual *flux-density* distribution in the air gap.

(b) *M.M.F. base*

With this base, the base rotor current is defined as that giving a (square-wave) m.m.f. per pole which is equal to that of the 'flat-topped armature-reaction' wave caused by balanced 3-phase armature currents of a unit peak value. The flat-topped armature-reaction m.m.f. is discussed in Section 3.2.3.

(c) *Unit-voltage base*

In this case, the base rotor current is defined as that which, in the absence of saturation, would induce the rated voltage in the armature when open-circuited. This is akin to Stephen's definition of unit m.m.f.

(d) *Equal-mutuals base*

As the name implies, the base rotor currents in this case are chosen so that, for a machine with three circuits on the direct axis, all the per-unit mutual reactances (on the direct axis) are equal.

Rankin shows, with the aid of his design equations, that these four definitions give numerically different per-unit impedances for the same machine. To emphasise this important point, his numerical example is included here. Consider a machine with the following design constants:

$$K_\phi = 0·98$$

$$\frac{4}{\pi}\frac{A_{d1}}{F_{d1}} = 1·10$$

$$D_{d1n} = 0·84$$

$$D_{don}y_{nd} = 0·73$$

$$X_{ad} = 1·20$$

The numerical values of X_d, X_{afd} and X_{ff} for the different bases, expressed as percentages of the values for the X_{ad} base, are given in Table 1.

Table 1 *Calculated numerical values of X_d, X_{afd} and X_{ff} (percentages of values for the X_{ad} base)*

	X_d	X_{afd}	X_{ff}
X_{ad}	100	100	100
M.M.F.	100	91	83
Unit voltage	100	83	69
Equal mutuals	100	105	110

The considerable variation in these values, and the consequent need for complete consistency in the use of base currents, are evident.

Gibbs[6], in his treatment of basic unified machine theory, affords an example of a fairly common situation in which a per-unit system is briefly introduced, but it is oversimplified to the point of being strictly incomplete. In this system, the base voltage and current of each armature phase are defined (normally) as the rated values. The per-unit field voltage and current of a synchronous machine are subsequently used, but not explicitly defined. One might reasonably assume that the base field quantities should be defined similarly as 'the rated values' (i.e. those corresponding to rated armature output at rated speed), and indeed this is a not uncommon misconception of principle among students. If this were so, however, it would constitute yet another definition to be added to the main ones discussed by Rankin and, moreover, would be one having undesirable mathematical properties, as discussed in Sections 2.1 and 2.2.

White and Woodson[7], in their approach to unified machine theory, choose for rotor circuits an 'equal-mutuals' definition (in contrast with Rankin's preference), but they mention without detail that there are alternatives. Their definition, which is applied on both the direct and the quadrature axes, is assumed without qualification to apply for *any* number of circuits on each axis. However, it is important to note (as will be discussed in Section 3.2.4) that it is impossible in principle to realise such a per-unit system, unless a particular, and generally unacceptable, restriction on the nature of the machine circuits is assumed.

In presenting the fundamental theory of machines, simplifying assumptions are often legitimately made. Thus Adkins[8], having assumed that the same shape of m.m.f. wave is produced by all windings in a synchronous machine, is able to define unit rotor current as producing 'the same m.m.f. as rated 3-phase armature currents'. With this simplification, however, the X_{ad} and m.m.f. base definitions are identical, and both are equivalent to Adkins's definition. It may be noted that, in a practical machine, this principle of definition (i.e. basically by comparison of m.m.f.s) has significant shortcomings. This question is fully discussed in Sections 3.2.1 and 3.2.2.

In later analysis by Adkins and in treatments by other authors, the approximation is made that, in some cases (depending on the physical disposition of the windings and core), mutual inductances on the direct axis may be *assumed* equal. It is important to distinguish this situation of approximate equality from that in which the mutual inductances are made exactly equal as by the use of Rankin's fourth definition.

Concordia[10], in his well known treatment of the synchronous machine, defines the base field current as that which establishes the same *space fundamental air-gap flux* as unit-peak 3-phase armature currents. This, as he mentions, is an alternative, and physically more helpful, definition of the X_{ad} base.

Rothe[11] gives a clear account of per-unit systems in single- and 3-phase circuit problems. The X_{ad} base is chosen for the synchronous machine with no discussion of alternatives, and it is stated that this base definition permits the performance equations to be represented by equivalent circuits in the main axes. However, other definitions equally permit this; indeed, as Rankin[2] states, a main advantage of the equal-mutuals base is the simplicity of its equivalent circuit. As is shown in Section 3.2.1, the most significant features of the X_{ad} base are that it is physically the most meaningful, and that the reactances in its equivalent circuit correspond to the (ordinary) reactances normally calculated by the designer.

Other authors writing on machines (e.g. References 12 and 13) and power-systems analysis (e.g. References 14 and 15) mention per-unit systems. In general, their brief treatments neither raise discussion of the problem of defining base currents in coupled secondary circuits, nor suggest that there are uncertainties either in this or in other matters.

In fact, there are troublesome variations in the treatment of a number of other matters. The 2-axis transformation is particularly important in this respect, giving rise to two distinct problems. First, different authors[8,10] derive identical per-unit 2-axis equations, while apparently defining their per-unit base magnitudes differently. In Section 3.1.1, this is shown to arise from a difference in the ordinary 2-axis transformation implicitly assumed, the transformation being physically correct in one case but not in the other.

The second problem is raised particularly clearly by Lewis[16], who discusses several matters which are treated in detail later. He recommends and adopts a physically correct 2-axis transformation with, however, a numerical factor of $\sqrt{(2/3)}$ rather than 2/3. A per-unit system is then defined so that the ordinary and per-unit equations are identical in form. While this may seem to be an attractive feature, it must be noted that the concept of the per-unit system is thereby fundamentally changed. There are important arguments against this choice; these are developed in Sections 3.1.1–3.1.3.

Following the above references to unified machine theory, the approach used by Kron, the originator of this theory, is of interest. In much of his work[9], he does not introduce per-unit systems. While arguments can be developed for and against this approach, it certainly has the merit of simplicity for a wide-ranging treatment of unified theory. Normalising the parameters of a circuit implies, in general, a change in its effective number of turns (see Chapter 2). This change must be allowed for mathematically, if in any theoretical development the circuit is interconnected with others. An example of this is a transformation of the impedance matrix of a primitive commutator machine, in which by applying a connection matrix the impedance equation of the *series*-connected machine is obtained.

Kron does refer to per-unit systems[17] however, and it is clearly his intention that the forms of any per-unit performance equations should be the same as those of the corresponding ordinary equations.†

Other variations arise from the different definitions of alternating voltage and current base quantities, sometimes as r.m.s. values, but more commonly (and often without explicit indication) as peak values. In some simple treatments[5, 12, 18] the matter is not raised. This variation is at first surprising, since r.m.s. values are generally assumed in transformer practice. Rothe[11], for example, uses r.m.s. for transformers, but peak values for machines, without justifying the difference. There are in fact good reasons for the change. The matter is discussed, with references, in Section 4.1.

Some simple recommendations[18] have been made on preferred practice for per-unit systems in a.c. machines. These define the rated voltage and current (by implication r.m.s.) per equivalent star phase, and total rated apparent power as the fundamental base magnitudes for the stator of a synchronous machine. Preferred rotor values are not

† For general interest, the reader is referred to a matter akin to, but distinct from, that of per-unit systems, namely Kron's use of the inductance tensor[17] as a 'metric tensor'. This may be viewed as a tensor of characteristic values, and equations of performance can be re-expressed in an interesting way by multiplication with the inverse of the metric tensor. The resultant equations are not, however, dimensionless.

discussed. Elsewhere[16], per-phase rated apparent power has been described as the fundamental base of power (though that view will not be agreed with here). Additionally, per-unit *line* values are sometimes defined[11], which clearly have a legitimate use in particular problems. In Sections 4.1 and 4.2 these alternatives (including peak and r.m.s.) are treated, not as being incorrect, but as being logically related to one fundamental set of base magnitudes. This relationship is given formal expression through the notion of 'associated' per-unit quantities, which proves to be a helpful concept.

Difficulties of understanding are often increased by having no notational distinction between per-unit and ordinary parameters. It is literally true that the per-unit equations of one author are identical to the ordinary equations of another, although equations which might be expected to correspond in fact differ. To avoid any such difficulties a distinct notation is employed here, italic type being used for ordinary parameters and bold-face type being used for per-unit parameters. This is the opposite arrangement to that chosen by Rankin, but it is the more reasonable.

2 BASIC ANALYSIS OF COUPLED CIRCUITS

2.1 Single-phase transformer

In this Section, the ordinary circuit and power equations of a 2-winding transformer are normalised, using arbitrarily different bases for each voltage and current. The resultant equations are examined, and it is demonstrated that, to obtain a useful per-unit system, it is necessary to make some bases dependent on other freely chosen values.

The ordinary circuit and power equations of the coupled windings shown in Fig 1. may be written in phasor form as

$$V_1 = Z_{11} I_1 + jX_{12} I_2 \qquad (2.1)$$

$$V_2 = jX_{12} I_1 + Z_{22} I_2 \qquad (2.2)$$

$$P_1 = \text{Re } V_1 I_1^* \qquad (2.3)$$

$$P_2 = \text{Re } V_2 I_2^* \qquad (2.4)$$

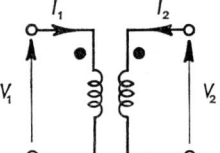

Fig. 1 Single-phase transformer

The assumption of sinusoidal excitation and the use of phasor notation avoid the need to introduce discussion of per-unit time at this stage. This is treated in Section 4.3.

The four variables of voltage and current may next be divided by four independently chosen base (or 'characteristic') values, so producing normalised variables. By making compensating adjustments to the coefficients (i.e. impedance terms) of eqns. 2.1 and 2.2, these equations become

$$\frac{V_1}{v_{10}} = \left(Z_{11} \frac{i_{10}}{v_{10}}\right)\left(\frac{I_1}{i_{10}}\right) + j\left(X_{12} \frac{i_{20}}{v_{10}}\right)\left(\frac{I_2}{i_{20}}\right) \qquad (2.5)$$

$$\frac{V_2}{v_{20}} = j\left(X_{12} \frac{i_{10}}{v_{20}}\right)\left(\frac{I_1}{i_{10}}\right) + \left(Z_{22} \frac{i_{20}}{v_{20}}\right)\left(\frac{I_2}{i_{20}}\right) \qquad (2.6)$$

Each bracketed group is dimensionless, and could be replaced by a per-unit symbol. It is clear, furthermore, that the impedance bases, for example v_{10}/i_{10} for primary impedance, are *dependent* quantities. However, the resulting per-unit system is too general to be useful; the reasons for this are examined next.

The reason most commonly cited is that the resulting per-unit mutual inductances would generally be asymmetric. This is taken to imply that the analysis degenerates into a meaningless array of numbers in which the form of the per-unit equations and the nature of the per-unit parameters do not resemble their ordinary counterparts. An inspection of eqns. 2.5 and 2.6 shows that symmetry would be assured by the following restricting relationship between the voltage and current bases:

$$\frac{i_{20}}{v_{10}} = \frac{i_{10}}{v_{20}}$$

or
$$v_{10}i_{10} = v_{20}i_{20} \qquad (2.7)$$

Although it is clear that, if the per-unit mutual reactances are not symmetric, they cannot be represented in per-unit equivalent circuits by simple passive components, this is not a compelling reason for requiring symmetry. However, further considerations, particularly of the equations for power, yield conclusive arguments in favour of maintaining symmetry. Thus, if the voltages and currents of eqns. 2.3 and 2.4 are normalised to the same four independent bases, the power equations can be rewritten in dimensionless groups as

$$\frac{P_1}{v_{10}i_{10}} = \mathrm{Re}\left(\frac{V_1}{v_{10}}\right)\left(\frac{I_1^*}{i_{10}}\right) \qquad (2.8)$$

$$\frac{P_2}{v_{20}i_{20}} = \mathrm{Re}\left(\frac{V_2}{v_{20}}\right)\left(\frac{I_2^*}{i_{20}}\right) \qquad (2.9)$$

It is now clear that the bases of power are dependent quantities, and will moreover be different for the primary and secondary circuits unless $v_{10}i_{10} = v_{20}i_{20}$—the condition of eqn. 2.7. If they are allowed to differ, it follows that per-unit components of power (both loss and throughput) do not have the same relative proportions as do their ordinary counterparts. Per-unit power calculated in this way (and equally per-unit energy) is not 'conserved', so that the concept of efficiency, for example, would have no value in such a per-unit system.

We now see more clearly that a loss of reciprocity in the mutual-inductance term does indeed indicate that the form of the analysis has degenerated; the behaviour of the per-unit variables does not have a fundamental similarity to the behaviour of the corresponding ordinary variables. The restriction of eqn. 2.7 is therefore seen to be necessary, and can be enlarged to define the base of power for this example. Thus:

$$p_0 = v_{10}i_{10} = v_{20}i_{20} \tag{2.10}$$

It must, however, be noted that it is not true of every useful per-unit system that the base of power is equal to the product of the voltage and current bases in any circuit. The preceding arguments are not affected by introducing a proportionality constant; thus $p_0 = Kv_0i_0$ as long as K is the same for all circuits. A common per-unit system for simple 3-phase system analysis, where the base voltages and currents are the *per-phase* r.m.s. rated values but the power base is the corresponding *total* mean, affords an example where $K = 3$. However in many instances, and the 2-axis variables of the per-unit system discussed in Sections 3.1.1 and 4.1 are examples, K is arranged to be unity. The essential principle involved is that of eqn. 2.7, which requires that the voltage–current base product must be single-valued.

An alternative expression of eqn. 2.7 is worth noting; namely

$$\frac{i_{20}}{i_{10}} = \frac{v_{10}}{v_{20}} = N \tag{2.11}$$

where N may be freely chosen. It is clear that in the per-unit system (as compared with the ordinary system) the ratio of primary to secondary currents is changed by a factor N, and the ratio of voltages is changed by a factor $1/N$. The effect is identical to that of replacing the secondary variables by variables referred to the primary winding, through an ideal transformer of primary/secondary turns ratio N. An important benefit for the per-unit system follows by making N equal to 'the turns ratio' between the actual windings; and this step is usually made in any preferred per-unit system. The advantages emerge clearly in later Sections. They arise from the fact that, by this step, the concepts of the turns ratio and referred values are effectively absorbed by the per-unit system, and are thus eliminated. (A per-unit systems analysis not incorporating this step has however been proposed[37].)

It is appropriate to note at this stage that many of the uncertainties of per-unit systems arise from the fact that it is not possible to define uniquely 'the turns ratio' between circuits. Considerations which lead to particular choices of N are discussed in Section 2.3 (mainly for the transformer), and Sections 3.2 and 3.3 (for rotating machines).

Having chosen the voltage and current bases of one circuit independently, one further quantity may be chosen independently for the other circuit; i.e. the base of voltage or current, or the turns ratio. The other two quantities follow by dependence, as do those of power and impedance. It is for this reason that the choice of a secondary-current base is said[2] to be equivalent to the choice of a turns ratio. Moreover, assuming for the moment that the turns ratio of this simple transformer has been chosen as the actual ratio of wound turns, then from eqn. 2.11 $Ni_{10} = i_{20}$; which implies that unit current in either winding produces equal m.m.f. around the transformer core.

This affords a very simple example of a principle which will recur throughout this Monograph, sometimes in extended and much less obvious forms (see Section 3.1.2). This may be referred to as the principle of *equal effect*, and proves to be valuable as an aid to physical interpretation of the per-unit system. In the present case, the 'effect' is the production of m.m.f. round the core of the transformer, and it may be said that a unit current in either the primary or the secondary winding is of equal effect in this sense.

Substituting from eqn. 2.11 in eqns. 2.5 and 2.6, we obtain

$$\left(\frac{V_1}{v_{10}}\right) = \left(Z_{11}\frac{i_{10}}{v_{10}}\right)\left(\frac{I_1}{i_{10}}\right) + j\left(NX_{12}\frac{i_{10}}{v_{10}}\right)\left(\frac{I_2}{i_{20}}\right) \quad (2.12)$$

$$\left(\frac{V_2}{v_{20}}\right) = j\left(\frac{X_{12}}{N}\frac{i_{20}}{v_{20}}\right)\left(\frac{I_1}{i_{10}}\right) + \left(Z_{22}\frac{i_{20}}{v_{20}}\right)\left(\frac{I_2}{i_{20}}\right) \quad (2.13)$$

It is clear that the impedance bases are dependent quantities which are defined as follows:

$$\left.\begin{array}{l} Z_{10} = \dfrac{v_{10}}{i_{10}} \\[1em] Z_{20} = \dfrac{v_{20}}{i_{20}} \end{array}\right\} \quad (2.14)$$

From eqns. 2.11–2.14, it is possible to summarise some properties of the per-unit parameters, as follows:

$$\begin{aligned}
\boldsymbol{V}_1 &= \frac{v_1}{v_{10}} = \left(\frac{V_1}{N}\right)\frac{1}{v_{20}} \\
\boldsymbol{I}_1 &= \frac{I_1}{i_{10}} = (NI_1)\frac{1}{i_{20}} \\
\boldsymbol{V}_2 &= \frac{V_2}{v_{20}} = (NV_2)\frac{1}{v_{10}} \\
\boldsymbol{I}_2 &= \frac{I_2}{i_{20}} = \left(\frac{I_2}{N}\right)\frac{1}{i_{10}} \\
\boldsymbol{Z}_{11} &= \frac{Z_{11}}{Z_{10}} = \left(\frac{Z_{11}}{N^2}\right)\frac{1}{Z_{20}} \\
\boldsymbol{Z}_{22} &= \frac{Z_{22}}{Z_{20}} = (N^2 Z_{22})\frac{1}{Z_{10}} \\
\boldsymbol{X}_{12} &= \left(\frac{X_{12}}{N}\right)\frac{1}{Z_{20}} = (NX_{12})\frac{1}{Z_{10}}
\end{aligned} \qquad (2.15)$$

Each of eqns. 2.15 demonstrates the same important result: any per-unit parameter has the same value, whether it is evaluated by referring (using turns ratio N) the corresponding ordinary parameter to the primary winding, and normalising it to the appropriate primary base magnitude, or by referring it to the secondary winding and normalising it to the corresponding secondary base magnitude. In this connection, it may be observed that 'referred mutual reactance' (i.e. NX_{12} referred to the primary winding, X_{12}/N referred to the secondary) is perhaps less familiar than the other referred parameters. It happens that the need to make this reference can be avoided simply by normalising X_{12} to the current base of one circuit, and to the voltage base of the other; i.e.

$$\boldsymbol{X}_{12} = X_{12}\left(\frac{i_{20}}{v_{10}}\right) = X_{12}\left(\frac{i_{10}}{v_{20}}\right) \qquad (2.16)$$

Some authors[16] use the result of eqn. 2.16 to define a separate base of mutual reactance equal to v_{10}/i_{20} or v_{20}/i_{10} [another variant is

$\sqrt{(v_{10}v_{20}/i_{10}i_{20})}$]. These approaches may be helpful, but they do single out mutual reactance for unnecessarily special treatment, whereas eqns. 2.15 treat it uniformly with all other circuit parameters.

Introducing per-unit symbols through eqns. 2.12 and 2.13,

$$V_1 = Z_{11}I_1 + jX_{12}I_2 \qquad (2.17)$$
$$V_2 = jX_{12}I_1 + Z_{22}I_2 \qquad (2.18)$$

and these equations are represented by the per-unit equivalent circuit of Fig. 2. It is meaningless to inquire whether this circuit is referred to

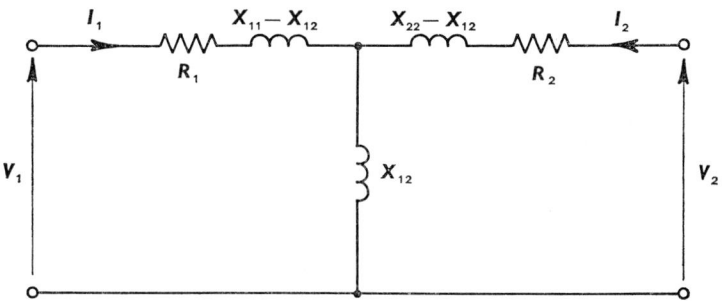

Fig. 2 Per-unit equivalent circuit of single-phase transformer

the primary or secondary winding, since, as the preceding discussion has shown, per-unit parameters are independent of this consideration. The corresponding ordinary values, referred to the primary or secondary side, are obtained by multiplying the per-unit value by its primary or secondary base, respectively.

Finally, the per-unit form of eqns. 2.8 and 2.9 for power are:

$$P_1 = \text{Re } V_1 I_1^* \qquad (2.19)$$
$$P_2 = \text{Re } V_2 I_2^* \qquad (2.20)$$

It has not been the primary purpose of this Section to establish a per-unit system for this example of a simple transformer, but rather to examine some fundamental ideas which are most simply demonstrated for this case. However, for the sake of completeness, it should be noted that, for the analysis of this device alone in steady-state sinusoidal operation,† v_{10} and i_{10} would normally be defined as the rated r.m.s.

† As opposed to the analysis of a larger system, of which this might form one element, where rated values would be defined with respect to the whole system.

values, with p_0 then following by dependence as the rated apparent power. v_{20} would probably be defined as the rated secondary value and i_{20} follows by dependence, but Section 2.3 should be read in this connection.

2.2 Generalisation to multiple circuits

In this Section, the ideas introduced in Section 2.1 are extended to cover several intercoupled circuits, allowing for relative motion between them, and so covering the representation of rotating machines.

Consider the voltage/current matrix equation of ordinary instantaneous parameters representing n intercoupled circuits:

$$[v] = [Z][i] \qquad (2.21)$$

where $[Z]$ contains motional terms, and has the general form

$$[Z] = [R] + [pL] \qquad (2.22)$$

which may be written as

$$[Z] = [R] + [Gp\theta] + [Lp] \qquad (2.23)$$

which is particularly of interest in pseudostationary reference frames.

Eqn. 2.21 can be normalised simply, without disturbing its mathematical form, by dividing all voltages and all currents by the base quantities of one circuit, say v_{10} and i_{10}. All elements of the impedance matrix can be normalised in a compensating fashion (i.e. to the dependent base v_{10}/i_{10}) to yield

$$\left[\left(\frac{1}{v_{10}}\right) \times v\right] = \left[\left(\frac{i_{10}}{v_{10}}\right) \times Z\right]\left[\left(\frac{1}{i_{10}}\right) \times i\right] \qquad (2.24)$$

This can be written
$$[v'] = [Z'][i'] \qquad (2.25)$$

where the prime indicates that a single set of base quantities has been employed. A different current base can now be incorporated for each circuit by the transformation

$$[i'] = [C][i] \qquad (2.26)$$

where in the present case

$$[C] = \begin{bmatrix} 1 & & & \\ & N_2 & & \\ & & N_j & \\ & & & N_n \end{bmatrix} \qquad (2.27)$$

This expression has been called the 'turns-ratio matrix'[17]. It implies new current bases for all the circuits except the primary:

$$i_{j0} = N_j i_{10} \qquad (2.28)$$

To maintain the invariance of total instantaneous electrical input power, it is known[6] that the voltage transformation automatically implied by the current transformation of eqn. 2.26 is

$$[v] = [C_t][v'] \qquad (2.29)$$

The significance of this power-invariant transformation of reference frame, represented jointly by eqns. 2.26 and 2.29, merits some discussion both for the present purpose and also in preparation for its use in Section 3.1. Essentially, the joint use of these two equations ensures the following properties in the new reference frame relative to the old:

(a) Terms in the impedance matrix having the mutual-inductance form are reciprocal.

(b) Not only is the total electrical input power unchanged, but also division between different uses is unchanged. Thus the new reference frame embodies the same mechanical power, total ohmic-power loss and rate of change of total stored magnetic energy. Consequently, the efficiency is unchanged in the new reference frame, and since the shaft speed is the same, the torque is also unchanged.

(c) The methods of calculating all the above (i.e. the form of the mathematics) have the same nature in both reference frames.

In short, therefore, all the important physical features of the system are preserved in precisely the manner which must be required of an acceptable per-unit system. Accordingly, eqn. 2.29 must be accepted as an automatic consequence of eqn. 2.26, and further, since $[C]$ has the particular form of eqn. 2.27 and $[C_t] = [C]$ in this case, it follows that eqn. 2.29 implies new voltage bases:

$$v_{j0} = \left(\frac{1}{N_j}\right) v_{10} \qquad (2.30)$$

The simple conclusion is thus a generalisation of that in the previous Section; namely that *in each circuit which is represented in the per-unit* system, *the voltage–current base product must be the same.* Thus eqns. 2.28 and 2.30 give

$$v_{j0} i_{j0} = v_{10} i_{10} \qquad (2.31)$$

It is also known that the equation for the transformed impedance matrix, corresponding to the current and voltage transformation equations (eqns. 2.26 and 2.29, respectively), is

$$[Z] = [C_t][Z'][C] \qquad (2.32)$$

In the present problem, this gives

$$[Z] = \begin{bmatrix} 1 & & & & \\ & N_2 & & & \\ & & N_j & & \\ & & & N_n \end{bmatrix} \times \begin{bmatrix} Z'_{11} & Z'_{12} & & Z'_{1j} & \\ Z'_{21} & Z'_{22} & & & \\ Z'_{j1} & & & Z'_{jj} & \\ & & & & Z'_{nn} \end{bmatrix} \times \cdots$$

$$\cdots \begin{bmatrix} 1 & & & \\ & N_2 & & \\ & & N_j & \\ & & & N_n \end{bmatrix} \qquad (2.33)$$

Note that, since eqns. 2.26 and 2.29 have the physical interpretation of referring the jth circuit parameters to the primary circuit through a primary/secondary turns ratio N_j, we may expect a generalised result equivalent to that obtained in Section 2.1; namely that any per-unit parameter of the jth circuit can be evaluated, either by normalising the ordinary parameter to the base quantity of that circuit, or by referring it to any other (kth) circuit and normalising it to the corresponding kth base quantity. Thus,

$$\left. \begin{aligned} v_j &= \left(\frac{v_j}{v_{jo}}\right) = \left(\frac{N_j}{N_k}\right)\left(\frac{v_j}{v_{ko}}\right) \\ i_j &= \left(\frac{i_j}{i_{jo}}\right) = \left(\frac{N_k}{N_j}\right)\left(\frac{i_j}{i_{ko}}\right) \\ Z_{jj} &= \frac{Z_{jj}}{Z_{jo}} = \left(\frac{N_j}{N_k}\right)^2 \left(\frac{Z_{jj}}{Z_{ko}}\right) \\ Z_{kj} &= \left(\frac{N_k}{N_j}\right)\left(\frac{Z_{kj}}{Z_{jo}}\right) = \left(\frac{N_j}{N_k}\right)\left(\frac{Z_{kj}}{Z_{ko}}\right) \end{aligned} \right\} \qquad (2.34)$$

Eqns. 2.34 are seen to be a generalisation of eqns. 2.15 in Section 2.1. They are obtained by evaluating the elements of $[\mathbf{Z}]$ from eqn. 2.33, and applying the results of eqns. 2.28 and 2.30, together with the equation for the dependent impedance base of the jth circuit:

$$Z_{jo} = \frac{v_{jo}}{i_{jo}} \qquad (2.35)$$

The discussion of this Section may appear to labour with what is, after all, a common feature of amost all per-unit systems in use, but it has provided a fundamental insight into the need, expressed by eqn. 2.31, for a single voltage–current base product throughout all circuits. For example, it is clearly not permissible in any useful per-unit system to define base voltage and current as 'the rated values' in every circuit of a device, since, in general, their products would not be single-valued.

Further, it is helpful to have examined this matter closely before the introduction of a small but important variation of principle. It should be noted that we have explored the requirement that all circuits representing a device in one particular reference frame should share a single voltage–current base product. In particular, for the voltage/current per-unit equation

$$[v] = [\mathbf{Z}][i] \qquad (2.36)$$

every circuit represented in $[\mathbf{Z}]$ must satisfy this requirement. If the reference frame is transformed, so that the same device is represented by

$$[v''] = [\mathbf{Z}''][i''] \qquad (2.37)$$

nothing in the foregoing theory demands that the circuits represented in $[\mathbf{Z}'']$ must have the same base product as those in $[\mathbf{Z}]$; the base products in $[\mathbf{Z}]$ and $[\mathbf{Z}'']$, respectively, must be single-valued, but the two values can differ. It happens that this permissible flexibility can in some instances offer a considerable advantage for the per-unit system used in a transformed reference frame. The recommended 3-phase/2-axis per-unit system proves to be an important example. On the basis of the above discussion, it becomes possible to place on a sound theoretical basis the relationship between this long-established per-unit system and the more recent notions of power invariance, which at first sight appear to be in conflict.

It is most desirable that all the steps taken in setting up a particular per-unit system should be kept separate so that the possibility of confusion is avoided. This has been done so far, and is a feature in following Sections. The four main and distinct steps which are advocated are:
   (i) assembly of all the ordinary circuit equations representing a device
   (ii) application of any transformation of a reference frame which may be required for analytical convenience—the 'turns ratio' transformation is not included here
   (iii) normalisation to a single set of base quantities of all parameters within one reference frame
   (iv) introduction of appropriate turns ratios, in accordance with principles which are discussed in Sections 2.3, 3.1.2 and 3.2.

It is desirable to conclude this Section with a brief comment on the physical interpretation of the results of power-invariant transformations in general. The manner in which important physical features of the electromechanical system or device are preserved has already been discussed. From this, one may be led to suppose, since the transformed system has the same physical nature as the original in so many respects, that it might be possible to interpret it in terms of some new arrangement of coupled circuits—in fact, to construct a simple physical model of the transformed system. Particular cases where this is so are frequently discussed in the literature, but in fact, while the broad implications of a particular mathematical transformation are generally quite clear, it is not always the case that a straightforward physical model of the transformed system can be identified. In general, it is necessary to perform the transformation, and then to examine the resulting equations to determine what physical interpretation is possible.

This matter arises in several Sections of Chapter 3, since the assessment of the value of a particular per-unit system inevitably involves discussion of the physical nature of the device under analysis. Where the analysis involves a transformation of the reference frame, it is, of course, necessary to understand the physical implications of this transformation.

2.3 Turns ratio and leakage reactance

In the preceding Sections, the value of the turns ratio N has been taken to be arbitrary, and it is now necessary to discuss the basis on which it can be most meaningfully chosen for a per-unit system in practice. Several authors discuss alternative turns ratios[11,13,16], but they are only concerned with describing the associated equivalent-circuit configurations which can result. A fuller treatment follows, and serves as an introduction to Section 3.2.

The actual turns ratio might appear an obvious choice for the transformer, and it is generally accepted that, with this choice, the two series reactances of Fig. 2 are the per-unit 'leakage' reactances. A feature of the per-unit system is also clear from Fig. 2; namely that the total self reactance of either circuit is equal to the sum of the mutual and corresponding series reactances without any need arising to refer reactances. Thus, for primary and secondary, respectively,

$$\left. \begin{aligned} X_{l1} &= \frac{1}{Z_{10}}(X_{11} - N X_{12}) \\ X_{l2} &= \frac{1}{Z_{20}}\left(X_{22} - \frac{X_{12}}{N}\right) \end{aligned} \right\} \quad (2.38)$$

where N is taken as the actual ratio. However, in practice, the nominal voltage ratio is perhaps the most common choice in the literature for the per-unit circuit analysis of the power transformer, since it has the advantage that unit voltage becomes exactly the rated value in both primary and secondary circuits. Other choices are sometimes made for various analyses, as discussed later in this Section. None of these is essentially more correct than the others, but it may be said that for many power transformers, because of the tightness of the coupling, the differences between these turns ratios is small, and the differences between the sums of the associated series reactances are then usually negligible. (Tap changing transformers cause obvious complications[37].)

For a machine, the distribution of coupled circuits is generally more complex, and two difficulties become apparent. First, the nominal voltage ratio cannot generally be chosen since it is often nonexistent, as for example with the permanently short-circuited damper windings of a synchronous machine. Secondly, it becomes important to recognise that there is, in general, no definition of the turns ratio which invariably leads to physically significant 'leakage' reactances, although in practice a turns ratio which does have this property can frequently be identified. The latter situation also exists in principle for the transformer, but because of the tight coupling and the simple nature of the field, it is not so apparent or of the same practical importance. Since the turns ratio and 'leakage' reactance commonly occur in design calculations, it is necessary to examine carefully, beginning with the simpler case of the transformer, the assumptions on which the calculations are based. However, it is in the design calculations for machines (see, for example, Sections 3.2.1 and 3.2.2) that the outcome of this discussion has the most important effect.

A common approach in the design process is to identify a turns ratio between two windings, based on an analysis of the fields of mutual and leakage flux which link the windings. A turns ratio is then identified as

$$N = \frac{\text{total flux linkages of all mutual flux with primary winding}}{\text{total flux linkages of all mutual flux with secondary winding}} \quad (2.39)$$

Additionally, individual leakage inductances for each winding are calculated as

$$L_l = \frac{\text{total flux linkages of leakage flux of winding}}{\text{current in winding}} \quad (2.40)$$

It is normal procedure to identify the reactance equivalents of these individual leakage inductances referred to one winding with the series reactances of an ordinary equivalent circuit referred to one winding, having the general form of Fig. 2; it being understood that the turns ratio used for reference is that given by eqn. 2.39.

A procedure of this nature is so common in design work that it may easily be supposed that any problem of inductively coupled coils can be analysed in the same terms. It is important therefore to stress that this approach lacks general validity, and is only justified for circumstances in which the nature of the fields linking the coils can be recognised to be special in the sense discussed below. More generally, two difficulties arise.

First, the turns ratio defined by eqn. 2.39 is not a single-valued constant, but is dependent on the load condition of the coupled coils; i.e. on the relative values of the currents flowing in them. It follows that, for the representation of the coils over a range of load conditions, any particular value of N determined by eqn. 2.39 from a particular field plot totally lacks significance.

Secondly, it is not generally possible to relate the values of the series elements of the equivalent circuit to the leakage inductances calculated according to eqn. 2.40. Consequently, the value of the concept of individual leakage inductances is virtually lost, since it depends on their simple relationship to the equivalent circuit. It must be emphasised that the problem is not simply one of finding an appropriate turns ratio by which to refer the equivalent circuit to one winding, though this would be serious enough in view of its variability. The more fundamental difficulty is that there is no direct relationship, in general, between the series reactances of the equivalent circuit and the reactances derived through eqn. 2.40. It is easy, for example, to postulate a simple field problem with no leakage flux for which the two series reactances of the equivalent circuit cannot be made simultaneously zero by any choice of turns ratio.

A sufficient restriction on the nature of the fields to ensure that both these difficulties disappear is that every tube of flux which is mutual between the windings may be considered to link both perfectly, thus forming an unvarying pattern of mutual flux density; i.e. there are negligible tubes of mutual flux which partially link the windings, as in Fig. 3, and the shape of the mutual-flux paths does not vary with load. Such a field pattern may be referred to as 'ideal'; the corresponding turns ratio given by eqn. 2.39 being the 'ideal ratio' N_i.

These points are demonstrated in Appendix 6.2, by considering a particular configuration of coupled coils; the coils are shown to lack ideal coupling in general, but to become ideally coupled in the restricted circumstances.

It is observed in many practical cases that the field pattern may be treated as ideal to a sufficiently good approximation, and important examples of this are discussed in Chapter 3. (However, some degree of approximation, although often slight, is always involved in the ideal-ratio concept.) This tends to happen, partly because of the constructional symmetry of many practical devices, and partly because of the use of high-permeability iron cores which concentrate and simplify the field patterns. In such cases, it is reason-

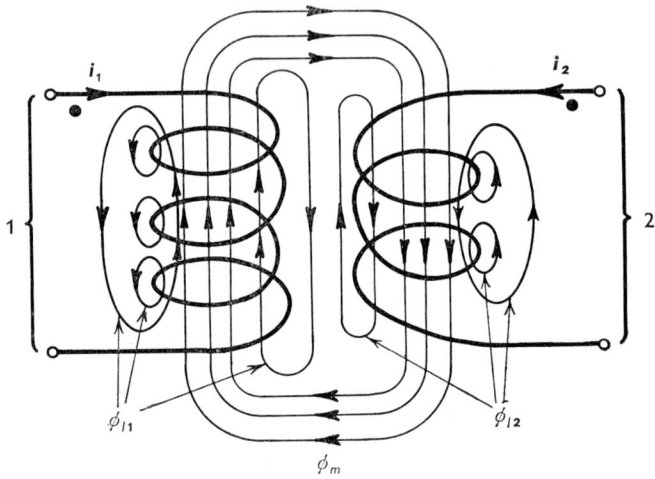

Fig. 3 Situation implied by the concept of 'ideal turns ratio'. The directions of currents and fluxes are consistent with Fig. 1

able for the designer recognising this property to base his calculations of individual-winding leakage reactances on a choice of turns ratio in accordance with eqn. 2.39. However, it must be remembered that such calculations inherently incorporate the particular turns ratio N_i and that subsequent circuit manipulations must not conflict with this.

It is common design practice to calculate the leakage reactance of a winding referred to itself, and then to refer this to another coupled winding by multiplication with the square of a turns ratio N. In per-unit terms, this implies

$$\left. \begin{array}{l} \boldsymbol{X}_{l1} = \dfrac{X'_{l1}}{Z_{10}} = \dfrac{X'_{l1}}{N^2 Z_{20}} \\[1em] \boldsymbol{X}_{l2} = \dfrac{X'_{l2}}{Z_{20}} = \dfrac{N^2 X'_{l2}}{Z_{10}} \end{array} \right\} \quad (2.41)$$

where the prime indicates that the leakage reactances are those calculated in design from the flux linkages. It is emphasised that the convenient form of eqns. 2.41 is only equivalent to the more general form of eqns. 2.38 when the flux pattern is recognisably ideal and the individual leakage reactances are calculated in accordance with that pattern. The turns ratio used for reference (N in eqns. 2.41) is the corresponding ideal ratio, N_i. If, possibly for convenience of circuit analysis, another choice of N is made, eqns. 2.41 represent an incorrect procedure and eqns 2.38 must be used.

This point, which may be overlooked, may be shown as follows. Consider, with reference to eqns. 2.38, that X_{11} is composed of a leakage reactance X'_{l1} plus a contribution X_1 from the flux which ideally links both windings, and treat X_{22} in a similar manner. The coupling coefficient calculated for the flux linkages of X_1 and X_2 is then unity, and

$$\left. \begin{array}{c} X_{12}^2 = X_1 X_2 \\ \dfrac{X_1}{X_2} = N_i^2 \end{array} \right\} \quad (2.42)$$

It follows that $X_1 = N_i X_{12}$ and $X_2 = X_{12}/N_i$, and these values substituted in eqns. 2.38 produce the form of eqns. 2.41.

In practice, there are broadly two circumstances in which flux linkages may be treated as ideal. First, in some cases it may be recognised that, to a sufficiently good approximation, the nature of the particular flux pattern is ideal. An example of this, relating to nested damper windings in synchronous machines, will be found in Sections 3.2.2 and 3.2.3. Secondly, it may be recognised that the flux pattern, although not strictly ideal, is so in respect of the flux which is fundamental to the operation of the device. In that case, the real windings can be replaced by equivalent windings which do have ideal coupling. The definition of the equivalent windings must be such that there is no change in the mutual flux linkages which are of fundamental importance. The replacement of distributed 3-phase windings by equivalent concentrated windings provides an example of this and is discussed in Section 3.2.1.

Turning now to a brief consideration of flux patterns other than the ideal, it may be felt that some extended definition of turns ratio exists which is in some sense of fundamental physical significance. This is not so. Lewis[16], for example, gives a definition of preferred turns ratio as implied by eqn. 2.39, but assuming the particular case of primary excitation. This is perhaps convenient, and it may be noted that, when the flux-density pattern is ideal, this definition leads to the ideal ratio. However, it cannot be considered fundamental in a more general sense, since it is double-valued—a different answer being obtained (as Appendix 6.2 shows) with secondary excitation. Of course no fundamental distinction exists between the primary and secondary windings.

The leakage reactance is sometimes defined in terms of stored field energy, and Billig[19] provides an example of this approach. His definition in terms of field vectors is attractively simple, but might be taken to imply a turns ratio which would be the ratio of secondary/primary currents with the secondary short-circuited.† This definition does not

---

† The basic approach in Section 7 of Reference 19 is to define the total stored energy with one winding excited, say the primary, and all others short-circuited. The instantaneous total energy storage is then stated in the form

$$\frac{1}{2} \sum_{1}^{n} \oint i_j \overline{A}_t . dl_j \quad \text{(A)}$$

where i_j = current in the jth circuit

$\overline{A}_t$ = total vector potential field due to all currents.

The contour integral is taken around the jth circuit. Eqn. A, or more strictly one equivalent to it, is used to define the total leakage inductance referred to the primary winding, and each term of the series defines the individual jth leakage inductance. Thus the individual primary-leakage inductance for a 2-circuit model is

$$\tfrac{1}{2} L_{l1} i_1^2 = \tfrac{1}{2} i_1 \oint \overline{A}_t . dl_1 \quad \text{(B)}$$

and $\overline{A}_t$ is the sum of A_1 and $\overline{A}_2$, the primary and secondary contributions. But $\oint \overline{A}_1 . dl_1$ is equal to the self-induced primary flux linkage which may be written as $L_{11} i_1$, and $\oint \overline{A}_2 . dl_1$ is equal to the primary flux linkage mutually coupled from the secondary, which is inherently negative, and may be written as $-L_{12} i_2$. Substituting in eqn. B we have

$$\tfrac{1}{2} L_{l1} i_1^2 = \tfrac{1}{2} L_{11} i_1^2 - \tfrac{1}{2} L_{12} i_1 i_2$$

whence

$$L_{l1} = L_{11} - \left(\frac{i_2}{i_1}\right) L_{12} \quad \text{(C)}$$

reduce to the ideal ratio when the flux pattern is ideal unless the magnetising current is negligible. Yet other definitions may be proposed, none of which can be claimed as fundamental, but which may be convenient in the circuit analysis of specific types of problem.

Expressed in terms of the base impedance of the primary side, the series reactances of the equivalent circuit of Fig. 2 are, by eqn. 2.38,

$$X_{l1} = \frac{1}{Z_{10}}(X_{11} - NX_{12})$$

$$X_{l2} = \frac{1}{Z_{10}}(N^2 X_{22} - NX_{12})$$

The subscript l is used, although in general it has been seen that these reactances will not equal the design values of leakage reactance unless $N = N_i$. On consideration of these expressions, four possibly convenient choices of N readily come to mind. First, if N is defined as equal to $\sqrt{(X_{11}/X_{22})}$, the per-unit 'leakage' reactances become equal. Secondly, if N is chosen equal to X_{11}/X_{12}, or thirdly, if it is chosen as X_{12}/X_{22}, the first or second of the series reactances, respectively, become zero, thus simplifying the equivalent circuit. The third choice is that made by Rothe[11], since it is the ratio V_1/V_2 given by measurements of the voltages during an open-circuit test, if the resistances of the windings are negligible. This is strictly true only for secondary excitation. When the primary winding is excited, the open-circuit voltage ratio V_1/V_2 is equal to X_{11}/X_{12}, which is the reciprocal of the second choice mentioned above. The two open-circuit tests only give the same result when the coupling coefficient is unity, i.e. when

$$X_{12}^2 = X_{11} X_{22}$$

Both choices do, however, give positive series elements. This matter is discussed again below. Finally, a fourth definition could take N as unity. This choice is widely used in the manipulation of electronic

Comparing eqn. C with eqn. 2.38, it is clear that this analysis may be considered to imply a turns ratio $N = i_2/i_1$.

It must be emphasised that this form of definition is only really satisfactory when the magnetising current is to be neglected; this restriction is made explicit in a subsequent Section of Reference 19. It is advisable to note, however, that the attractively simple form of eqn. B has no more general significance.

circuits, but for the purpose of the present discussion it must be considered too arbitrary.

In basic principle, there is an infinite range of choices corresponding to an infinite range of equivalent circuits, all having the general form of Fig. 2. As Lewis[16] shows—and this may readily be confirmed—all the reactance values in the equivalent circuit are positive if N lies in the range

$$\frac{X_{12}}{X_{22}} \leqslant N \leqslant \frac{X_{11}}{X_{12}} \qquad (2.43)$$

Outside this range, one of the two series reactances will become negative. While this may be notionally repugnant, it must be appreciated that these individual reactances cannot be determined experimentally and hence there is no physical problem. Mathematically, the range of N is infinite. Where an ideal ratio can be recognised between two windings, it is evident that the use of this ratio will result in both series reactances being calculated as positive, since they correspond to the physically separate positive leakage flux linkages.

It is shown below that, for closely coupled windings, there is comparatively little change in the sum of the two series reactances over the range of N which keeps them positive. It is worth noting that, as a consequence, the total per-unit leakage reactance may in practice be comparatively unambiguous, even if an ideal turns ratio is not recognisable: this assumes that a 'practical' choice of N would be one that caused both primary and secondary series reactances to be calculated as positive. It follows from eqns. 2.38 that the total per-unit 'leakage' reactance is

$$\boldsymbol{X}_l = \frac{X_{11}}{Z_{10}} \left(1 - 2N \frac{X_{12}}{X_{11}} + N^2 \frac{X_{22}}{X_{11}} \right) \qquad (2.44)$$

The limits over which N can be varied while retaining positive series elements are given in eqn. 2.43, and the corresponding extreme limits of variation of the total 'leakage' reactance $\boldsymbol{X}_l$ are found by substituting these values into eqn. 2.44, giving

$$\frac{X_{11}}{Z_{10}} \left(1 - \frac{X_{12}^2}{X_{11} X_{22}} \right) \quad \text{and} \quad \frac{X_{11}}{Z_{10}} \left(\frac{X_{11} X_{22}}{X_{12}^2} - 1 \right)$$

In terms of the coupling coefficient $k = X_{12}/\sqrt{(X_{11}X_{22})}$, these become

$$\frac{X_{11}}{Z_{10}}(1-k^2) \quad \text{and} \quad \frac{X_{11}}{Z_{10}}\left(\frac{1}{k^2}-1\right) \qquad (2.45)$$

respectively. Table 2 compares the terms $1-k^2$ and $(1/k^2)-1$ for various values of k corresponding to tightly coupled coils thus showing the range of variation of the total 'leakage' reactance for positive series elements as the tightness of coupling is varied. With a 99% coupling the range is 2%; this increases until with a 90% coupling the range is about 22·3%.

Table 2 *Comparison of terms $1-k^2$ and $(1/k^2)-1$ with k for four values of k*

k	$1-k^2$	$(1/k^2)-1$	variation, %†
0·99	0·0199	0·0203	2·0
0·98	0·0396	0·0413	14·2
0·95	0·0975	0·1080	10·5
0·90	0·1900	0·2346	22·3

† The 'variation' here is defined as the difference between the maximum and minimum values of 'leakage' reactance for positive series elements divided by the value of the total 'leakage' reactance under the condition that the two series reactances are equal. The latter value can easily be shown to be

$$2\frac{X_{11}}{Z_{10}}(1-k).$$

3 CIRCUIT ANALYSIS OF MACHINES

In Section 3.1 attention is directed to some general problems concerning the nature of transformations from one reference frame to another in per-unit parameters, and a recommended procedure is given. This leads naturally to Section 3.2, in which various definitions of secondary-current bases for the synchronous machine are examined. Finally, the ways in which these ideas may be applied to some types of machine other than the synchronous are explored in Section 3.3.

3.1 Transformations and principles
3.1.1 3-phase/2-axis transformation

One particular form of the transformation equations, relating per-unit 2-axis variables to per-unit 3-phase variables, is considerably more common in the literature[8,10,11] than any other. This form, which is the same for instantaneous voltages and currents, is as follows:

$$\begin{bmatrix} v_d \\ v_q \\ v_z \end{bmatrix} = \tfrac{2}{3} \begin{bmatrix} \cos\theta & \cos(\theta-120°) & \cos(\theta-240°) \\ -\sin\theta & -\sin(\theta-120°) & -\sin(\theta-240°) \\ \tfrac{1}{2} & \tfrac{1}{2} & \tfrac{1}{2} \end{bmatrix} \begin{bmatrix} v_a \\ v_b \\ v_c \end{bmatrix} \quad (3.1)$$

$$\begin{bmatrix} i_d \\ i_q \\ i_z \end{bmatrix} = \tfrac{2}{3} \begin{bmatrix} \cos\theta & \cos(\theta-120°) & \cos(\theta-240°) \\ -\sin\theta & -\sin(\theta-120°) & -\sin(\theta-240°) \\ \tfrac{1}{2} & \tfrac{1}{2} & \tfrac{1}{2} \end{bmatrix} \begin{bmatrix} i_a \\ i_b \\ i_c \end{bmatrix} \quad (3.2)$$

The inverses of eqns. 3.1 and 3.2 are:

$$\begin{bmatrix} v_a \\ v_b \\ v_c \end{bmatrix} = \begin{bmatrix} \cos\theta & -\sin\theta & 1 \\ \cos(\theta-120°) & -\sin(\theta-120°) & 1 \\ \cos(\theta-240°) & -\sin(\theta-240°) & 1 \end{bmatrix} \begin{bmatrix} v_d \\ v_q \\ v_z \end{bmatrix} \quad (3.3)$$

$$\begin{bmatrix} i_a \\ i_b \\ i_c \end{bmatrix} = \begin{bmatrix} \cos\theta & -\sin\theta & 1 \\ \cos(\theta-120°) & -\sin(\theta-120°) & 1 \\ \cos(\theta-240°) & -\sin(\theta-240°) & 1 \end{bmatrix} \begin{bmatrix} i_d \\ i_q \\ i_z \end{bmatrix} \quad (3.4)$$

A variation of sign appears for certain terms in some quoted forms of these equations, but this is of no importance since it only reflects the choice of the q axis as lagging or leading the d axis.

At this stage the multiplying constant in eqns. 3.1 and 3.2 is accepted as 2/3 since, with this choice, the peak amplitude of the phase-voltage phasors in balanced synchronous operation is thereby made equal to $\sqrt{(v_d^2+v_q^2)}$, and similarly for current. This somewhat abstract

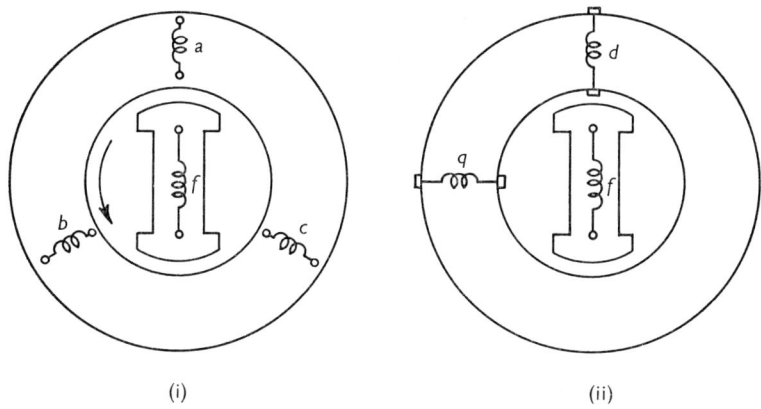

Fig. 4 Coil configurations
(i) The configuration of a, b and c coils related to the field winding, with relative motion between stator and rotor
(ii) The pseudostationary d and q coils which can replace the stator. The zero-sequence variables are treated separately

justification is expanded considerably later, but some mathematical detail is appropriate here. Under balanced conditions we may put

$$\left.\begin{aligned} v_a &= -\hat{v}\sin\omega t \\ v_b &= -\hat{v}\sin(\omega t - 120°) \\ v_c &= -\hat{v}\sin(\omega t - 240°) \end{aligned}\right\} \quad (3.5)$$

$$\theta = \omega t - \delta \quad (3.6)$$

where θ = electrical angle between axis of field winding and axis of stator phase a, and δ = torque angle.

Thus, for example, Fig. 4(i) defines the instant $t = 0$ when $\delta = 0$.

On substitution from eqns. 3.5 and 3.6 into eqn. 3.1, the following

37

results are readily obtained:

$$\left.\begin{array}{l} v_d = -\hat{v}\sin\delta \\ v_q = \hat{v}\cos\delta \\ v_z = 0 \end{array}\right\} \quad (3.7)$$

whence, as already noted
$$\hat{v} = \sqrt{(v_d^2 + v_q^2)} \quad (3.8)$$

which implies that the amplitude of the per-unit phase voltage is equal to the maximum per-unit v_d or v_q that may occur as δ is varied. Finally, a typical per-phase result for phase a may be noted from eqn. 3.3:

$$v_a = v_d \cos(\omega t - \delta) - v_q \sin(\omega t - \delta) \quad (3.9)$$

or, in complex phasor notation

$$\boldsymbol{V}_a = \boldsymbol{V}_d + \boldsymbol{V}_q \quad (3.10)$$

where
$$\left.\begin{array}{l} |\boldsymbol{V}_d| = v_d \\ |\boldsymbol{V}_q| = v_q \end{array}\right\} \quad (3.11)$$

The corresponding results for current are

$$\boldsymbol{I}_a = \boldsymbol{I}_d + \boldsymbol{I}_q \quad (3.12)$$

$$\left.\begin{array}{l} |\boldsymbol{I}_d| = i_d \\ |\boldsymbol{I}_q| = i_q \end{array}\right\} \quad (3.13)$$

It may be noted that transformations have been proposed which give results differing from those in eqns. 3.1–3.13, and the relative merits of these different approaches are discussed later in this Section. The next step, however, is to review some variations which occur in the literature of the manner by which eqns. 3.1 to 3.13 are derived and justified.

The manner in which Concordia[10] establishes the per-unit 2-axis parameters is the most common. He defines transformations for ordinary voltages and currents which have the same mathematical form as eqns. 3.1 and 3.2. This leads (for reasons that will be discussed) to nonreciprocal stator/rotor mutual inductances, but it is found by inspection that reciprocity is restored if all rotor currents are multiplied by 2/3. It is stated that this step is less drastic than it might appear, since the parameters are to be normalised anyway. Although the ultimate per-unit equations thus take a rational form, the procedure is particularly unsatisfying because of its lack of physical significance.

Adkins's[8] approach is essentially different, and it can be demonstrated to be physically far more significant, although this is not perhaps immediately obvious. His transformation is performed in per-unit variables, and his 2-axis current base is made equal to $\frac{3}{2}i_{ao}$. This contrasts with Concordia, whose current base equals i_{ao}, but the ultimate forms of their per-unit 2-axis equations are the same.

Adkins mentions a 2-axis machine model in his analysis, in which the d (direct-axis) and q (quadrature-axis) coils have the same number of turns as the phase coils of the actual machine. In contrast, as previously mentioned, Concordia's analysis is obscure in physical terms, and is in fact incapable of direct physical interpretation. This follows, as will be shown, from the physical incompatibility of his chosen ordinary voltage and current transformations. The ultimate justification for his analysis is that it yields the same set of per-unit equations as is obtained in the following rigorous manner.

The approach used here effectively enlarges Adkins's work by introducing an *initial transformation in ordinary parameters which is power-invariant*. It is then demonstrated that the d and q variables correspond to a particular physical model. The subsequent normalisation process is thus completely separated from the transformation process, so that the derivation of the per-unit system can be considered in the light of the physical model. The usefulness of the model concept is illustrated by showing that (for example) expressions for per-unit impedances can be interpreted in terms of it, and can be written down with little further mathematics.

In a discussion of the per-unit system, the d and q variables are of central importance. The z (zero-sequence) variables are not of great importance, because the per-unit system is defined so as to relate most simply to the operation of the machine under rated load (i.e normal balanced conditions) for which the zero-sequence variables are zero. However, some comment on the z variables is included here, although the per-unit treatment of symmetrical components as a whole is briefly discussed in Section 3.1.4.

The form of transformation for voltages given by eqn. 3.1 is reconsidered in terms of ordinary voltages:

$$\begin{bmatrix} v_d \\ v_q \\ v_z \end{bmatrix} = \tfrac{2}{3} \begin{bmatrix} \cos\theta & \cos(\theta-120°) & \cos(\theta-240°) \\ -\sin\theta & -\sin(\theta-120°) & -\sin(\theta-240°) \\ \tfrac{1}{2} & \tfrac{1}{2} & \tfrac{1}{2} \end{bmatrix} \begin{bmatrix} v_a \\ v_b \\ v_c \end{bmatrix} \quad (3.14)$$

This, of course, represents a transformation which can be written in matrix notation as

$$[v_{d,q,z}] = [C][v_{a,b,c}] \tag{3.15}$$

and, as indicated in Section 2.2, the transformation automatically implied for the currents in order to maintain power invariance is

$$[i_{a,b,c}] = [C_t][i_{d,q,z}] \tag{3.16}$$

or

$$\begin{bmatrix} i_a \\ i_b \\ i_c \end{bmatrix} = \tfrac{2}{3} \begin{bmatrix} \cos\theta & -\sin\theta & \tfrac{1}{2} \\ \cos(\theta-120°) & -\sin(\theta-120°) & \tfrac{1}{2} \\ \cos(\theta-240°) & -\sin(\theta-240°) & \tfrac{1}{2} \end{bmatrix} \begin{bmatrix} i_d \\ i_q \\ i_z \end{bmatrix} \tag{3.17}$$

The inverse of eqn. 3.14, which is identical in form to eqn. 3.3, should be noted:

$$\begin{bmatrix} v_a \\ v_b \\ v_c \end{bmatrix} = \begin{bmatrix} \cos\theta & -\sin\theta & 1 \\ \cos(\theta-120°) & -\sin(\theta-120°) & 1 \\ \cos(\theta-240°) & -\sin(\theta-240°) & 1 \end{bmatrix} \begin{bmatrix} v_d \\ v_q \\ v_z \end{bmatrix} \tag{3.18}$$

as should the inverse of eqn. 3.17 for current:

$$\begin{bmatrix} i_d \\ i_q \\ i_z \end{bmatrix} = \begin{bmatrix} \cos\theta & \cos(\theta-120°) & \cos(\theta-240°) \\ -\sin\theta & -\sin(\theta-120°) & -\sin(\theta-240°) \\ 1 & 1 & 1 \end{bmatrix} \begin{bmatrix} i_a \\ i_b \\ i_c \end{bmatrix} \tag{3.19}$$

Clearly, whereas the voltage-transformation equation (eqn. 3.14) has been chosen to have the preferred form of eqn. 3.1 immediately, the corresponding invariant-power current-transformation equation (eqn. 3.19) inevitably has a different form from that preferred in eqn. 3.2. First, the overall factor of $\tfrac{2}{3}$ is absent, and secondly the zero-sequence coefficients within the matrix are units instead of halves. The latter difficulty could be avoided by choosing these coefficients as $1/\sqrt{2}$ in the original forms; i.e. eqns. 3.1 and 3.2. The arguments for and against this choice are touched on in Section 3.1.4. However, the form of eqn. 3.2 is extremely common in the literature, and the present intention (as a first step) is simply to replace the units in the bottom row of eqn. 3.19 by halves, thus redefining i_z in a non-power-invariant manner. This results in the new equation:

$$\begin{bmatrix} i_d \\ i_q \\ i_z \end{bmatrix} = \begin{bmatrix} \cos\theta & \cos(\theta-120°) & \cos(\theta-240°) \\ -\sin\theta & -\sin(\theta-120°) & -\sin(\theta-240°) \\ \tfrac{1}{2} & \tfrac{1}{2} & \tfrac{1}{2} \end{bmatrix} \begin{bmatrix} i_a \\ i_b \\ i_c \end{bmatrix} \tag{3.20}$$

It must be emphasised that this is a sweeping procedure which in a more general transformation would produce physical and mathematical chaos. It is permissible here because of the particular nature of the problem in which the z variables are not 'coupled' to the d and q variables. This is familiar to the machines engineer (e.g. Reference 34, p. 418), and indeed represents one of the advantages of the 2-axis reference frame for analytical purposes. This is evident from the 2-axis impedance matrix, which is quoted in Appendix 6.1, eqn. 6.3. It is therefore possible to multiply throughout the zero-sequence equation by an arbitrary factor, as in eqn. 3.20, without disturbing the form of the direct- and quadrature-axis equations—the effects being limited, in fact, to the introduction of some anomalous coefficients in the equations for zero-sequence impedance and power, as noted later in this Section and in Section 4.1.

Considering now the ordinary d and q variables of eqns. 3.14 and 3.20, it is readily established that both the voltages and the currents are consistent with a '2-axis machine model', and that the model is one in which the pseudo-stationary d and q coils *each have the same number of effective turns as each of the a, b and c phase coils*. These ideas will perhaps be clarified by reference to Fig. 4; the a, b and c coils are replaced by the d and q coils, while no modification is implied either for the structure of any circuits on the rotor side, such as field coils or damper windings, or for any voltages or currents associated with them. This can best be seen by examining particular operating conditions of the machine, chosen so as to produce results from the transformation equations which can be easily interpreted. For the condition in which the field coil is excited by a time-varying current with the rotor stationary in the position of Fig. 4 (i.e. $\theta = 0$), $v_q = v_z = 0$. So from eqn. 3.18 $v_a = v_d$. This must imply that the a and d coils have equal turns. For the same position with the excitation applied to the a coil alone, $i_d = i_a$ from eqn. 3.20. For equality of the d axis m.m.f. in the two representations, the equality of turns is again implied.† Similar reasoning applies to the q coil, with the rotor turned through 90°. It is worth re-emphasising that the realisation of this proper 2-axis model depends on the use

† i_z is nonzero for this particular case, but it does not contribute to the d axis m.m.f., since it is magnetically uncoupled from the 2-axis model.

of the correct voltage and current transformations jointly observing power invariance.

The next step is the normalisation of the transformation equations. Eqn. 3.14 is normalised throughout to a base v_{ao}, the peak voltage per phase in the armature (see Section 4.1 for comparison of peak and r.m.s. values), when it immediately assumes the desired form of eqn. 3.1. In eqn. 3.20, however, an important principle must be introduced. This is that i_a, i_b and i_c are normalised to a base i_{ao}, the peak current per phase in the armature, but that i_d, i_q and i_z are normalised to a base $(3/2)i_{ao}$. This introduces a multiplying factor of 2/3 in the right-hand side of eqn. 3.20, and converts it to the same desired numerical form as eqn. 3.2.

Thus, eqn. 3.14 has been normalised throughout to one voltage base, but eqn. 3.20 has been normalised to a combination of two current bases; it will be realised that this introduces the permissible flexibility in the voltage–current base product, referred to in Section 2.2. Eqn. 3.20 relates the current variables in two different reference frames, and the essential requirement is that the voltage–current base product must be the same for all circuits in any one reference frame. The concept being introduced is that, for any circuits appearing in a per-unit analysis of the (a, b, c) reference frame, the base product must be $v_{ao}i_{ao}$, while in the (d, q, z) reference frame the base product must be $(3/2)v_{ao}i_{ao}$.

At this point the presence of the rotor circuits must be recalled. These circuits could have been represented throughout eqns. 3.1–3.20, but this would have served no useful purpose since the transformations specifically leave all rotor currents and voltages unchanged. However, in the normalisation process, the rotor circuits, which appear in a 2-axis analysis and are therefore part of the 2-axis reference frame, must conform to the same voltage–current base product as the other circuits of that reference frame.

Considering just the field winding, and noting that its treatment is typical of all rotor circuits, the ordinary voltage/current relationships in the 2-axis reference frame can be written as

$$[v_{d,q,z,f}] = [Z_{d,q,z,f}][i_{d,q,z,f}] \tag{3.21}$$

Normalising this equation throughout to the bases v_{ao} and $(3/2)i_{ao}$,

$$\left[\left(\frac{1}{v_{ao}}\right) \times v_{d,q,z,f}\right] = \left[\left(\frac{3i_{ao}}{2v_{ao}}\right) \times Z_{d,q,z,f}\right]\left[\left(\frac{2}{3i_{ao}}\right) \times i_{d,q,z,f}\right]$$

or
$$[v'_{d,q,z,f}] = [Z'_{d,q,z,f}][i'_{d,q,z,f}] \tag{3.22}$$

The prime again denotes the use of a single set of base quantities. A different base for each rotor current can now be introduced; this is equivalent to choosing turns ratios, as in Section 2.2. (The ways in which these can best be chosen for the per-unit system are discussed in Section 3.2.) Thus, if the field-current base is changed from $(3/2)i_{ao}$ to some value i_{fo}, the turns ratio is

$$N_f = i_{fo}/(3/2)i_{ao} \tag{3.23}$$

Before proceeding further it may be noted that, in some applications of transformation theory, the foregoing treatment would have led to unacceptable ambiguities. Its acceptability here is dependent on the nature of the particular analysis. Thus, were it considered desirable to make a per-unit analysis in the (a,b,c,f) reference frame, the (a,b,c) circuits would conform to a voltage–current base product $v_{ao}i_{ao}$, and the f circuit to a base product $(3/2)v_{ao}i_{ao}$. Consequently, the performance equations in this reference frame lack the desirable features (discussed in Section 2.2) preserved by a power-invariant transformation. To obviate this, it would be necessary to define new per-unit field variables conforming to a base-product $v_{ao}i_{ao}$, and the possibilities of confusion between the two definitions would be unacceptable. However, in practice, the (a,b,c,f) reference frame is hardly ever used since the inductance coefficients are θ-dependent[10,12].

As has been pointed out elsewhere[16], some detailed analyses have been made[21,22] in the (a,b,c,f) reference frame, but the view is taken here that these exceptions are insufficient to justify a revision of the per-unit system. While such analyses are unusual, one point should be made; it is quite common to quote a particular per-unit inductance parameter from the (a,b,c,f) reference frame; namely L_{afd}, the per-unit mutual inductance between the phase a and the field winding when phase a is on the direct axis. (The value between phase a and the various damper windings is also quoted.) This parameter may be considered as 'hybrid'; i.e. interrelating two reference frames whose voltage–current base products are unequal. In consequence, $\boldsymbol{L}_{afd}$ is non-reciprocal, and it may be evaluated in principle by

$$\boldsymbol{L}_{afd}\mathbf{p} = \left(\frac{i_{ao}}{v_{fo}}\right)(L_{afd}\mathrm{p}) \quad \text{or} \quad \boldsymbol{L}_{afd}\mathbf{p} = \left(\frac{i_{fo}}{v_{ao}}\right)(L_{afd}\mathrm{p})$$

In fact, where it is quoted[3, 8, 10], the latter form is always implicitly assumed. Thus,

$$\boldsymbol{L}_{afd}\mathbf{p} = \left(\frac{i_{fo}}{v_{ao}}\right)(L_{afd}\mathrm{p}) = N_f\left(\frac{3\,i_{ao}}{2v_{ao}}\right)(L_{afd}\mathrm{p}) \tag{3.24}$$

The foregoing discussion, of course, is not meant to imply that the (a, b, c) reference frame is not used in the analysis; this is far from true. Its normal use, however, is for a per-phase analysis in which the rotor circuits are not represented directly, but rather as effects referred to the stator, e.g. open-circuit voltage, transient and subtransient reactances etc. In this approach, clearly, no problem arises.

Returning to eqns. 3.21–3.23, the derivation of a number of per-unit impedances from these equations with the aid of physical ideas embodied in the 2-axis model is next demonstrated. In doing this the time will be normalised to a base of $1/\omega_o$, where ω_o is the rated electrical angular velocity. The idea of normalised time is implicitly assumed in eqn. 3.24; the merits of this will be considered in Section 4.3.

Considering the turns ratio N_f, the per-unit self impedance of the field winding must be

$$\boldsymbol{Z}_{ff} = \boldsymbol{R}_f + \boldsymbol{L}_{ff}\boldsymbol{p} = N_f^2\left(\frac{3i_{ao}}{2v_{ao}}\right)Z_{ff} \tag{3.25}$$

For the mutual inductance between the d and f coils, it should first be noted that the ordinary value must be L_{afd}, since it is known from the 2-axis model that the a and d coils have equal turns. Therefore

$$\boldsymbol{L}_{df}\mathbf{p} = N_f\left(\frac{3i_{ao}}{2v_{ao}}\right)(L_{afd}\mathrm{p}) = \boldsymbol{L}_{afd}\,\mathbf{p} \tag{3.26}$$

While the usefulness of the concept of the 2-axis model is clear, it is necessary to exercise care in its use. For example, although the a and d coils have equal turns they have different ordinary resistances. By equating i^2R losses in the two representations under conditions of balanced operation, say for an instant when

$$i_a = 1 \text{ A (peak)} \quad \text{and} \quad i_b = i_c = -0.5 \text{ A}$$

and from eqn. 3.20 $\sqrt{(i_d^2 + i_q^2)} = 1.5$ A

it is easily seen that $\{1^2 + (\tfrac{1}{2})^2 + (\tfrac{1}{2})^2\} R_a = (\tfrac{3}{2})^2 R_d$

Hence
$$R_d = R_q = (\tfrac{2}{3}) R_a \qquad (3.27)$$

whence
$$\boldsymbol{R_d} = \boldsymbol{R_q} = \left(\frac{3i_{ao}}{2v_{ao}}\right)\left(\frac{2}{3}\right) R_a = \left(\frac{i_{ao}}{v_{ao}}\right) R_a = \boldsymbol{R_a} \qquad (3.28)$$

Additionally, the ordinary self inductance of the d coil is different from that of the a coil, except, as may be seen from Appendix 6.1, in the particular case of the zero-sequence inductance being zero.† But L_{dd} can be conveniently expressed in terms of the synchronous inductance L_d per phase, which is, of course, defined for balanced steady-state operation. Considering the instantaneous current values to be the same as stipulated in the derivation of eqn. 3.27, and assuming these to flow in the phase coils in Fig. 4(i), it is evident that the total m.m.f. produced is instantaneously acting vertically and is 3/2 times that produced by the instantaneous current in phase a acting alone. Thus, the amplitude of the fundamental flux-density wave of the direct-axis armature reaction, as represented by L_d, is 3/2 times that produced by the peak phase current flowing in coil a alone, or, since it has the same number of turns, in coil d alone. It is thus immediately clear that

$$L_{dd} = (\tfrac{2}{3}) L_d \qquad (3.29)$$

whence
$$\boldsymbol{L_{dd}}\mathrm{p} = \left(\frac{3i_{ao}}{2v_{ao}}\right)\left(\frac{2}{3}\right) L_d \mathrm{p} = \left(\frac{i_{ao}}{v_{ao}}\right) L_d \mathrm{p} = \boldsymbol{L_d}\mathrm{p}$$

and therefore
$$\boldsymbol{L_{dd}} = \boldsymbol{L_d} \qquad (3.30)$$

† The problem of devising a (d, q, z) physical model which is fully equivalent to the 3-phase machine (even as idealised in transformation theory) is more substantial than has been suggested in this simple discussion, but it is not necessary to pursue it further. It can be noted, however, and for the present purpose this may be regarded as a mnemonic, that R_q, R_d and R_a are related in eqn. 3.27 as though the same volume of conductor material had been used to construct the d and q coils, as for the 3-phase coils a, b and c.

It may be noted that eqns. 3.25, 3.26, 3.28 and 3.30 agree with those quoted by Rankin[2], for example.

At this point, attention is specifically drawn to Appendix 6.1, which summarises the mathematical results of the transformation performed in accordance with eqns. 3.14 and 3.20, and of the subsequent normalisation process in eqn. 3.22. Some results given there overlap with those derived in this Section.

Some authors, notably Lewis[16], use a transformation for ordinary voltages and currents of a similar form to eqn. 3.14, but preceded by an overall factor of $\sqrt{(2/3)}$ instead of $2/3$. Additionally, the zero-sequence coefficients within the matrix are made equal to $1/\sqrt{2}$; this has already been noted as a possibility. Thus, the alternative transformation equations are

$$\begin{bmatrix} v_d \\ v_q \\ v_z \end{bmatrix} = \sqrt{\tfrac{2}{3}} \begin{bmatrix} \cos\theta & \cos(\theta-120°) & \cos(\theta-240°) \\ -\sin\theta & -\sin(\theta-120°) & -\sin(\theta-240°) \\ \sqrt{\tfrac{1}{2}} & \sqrt{\tfrac{1}{2}} & \sqrt{\tfrac{1}{2}} \end{bmatrix} \begin{bmatrix} v_a \\ v_b \\ v_c \end{bmatrix} \quad (3.31)$$

$$\begin{bmatrix} i_d \\ i_q \\ i_z \end{bmatrix} = \sqrt{\tfrac{2}{3}} \begin{bmatrix} \cos\theta & \cos(\theta-120°) & \cos(\theta-240°) \\ -\sin\theta & -\sin(\theta-120°) & -\sin(\theta-240°) \\ \sqrt{\tfrac{1}{2}} & \sqrt{\tfrac{1}{2}} & \sqrt{\tfrac{1}{2}} \end{bmatrix} \begin{bmatrix} i_a \\ i_b \\ i_c \end{bmatrix} \quad (3.32)$$

Such a transformation is well known to be power-invariant, and the particular choice of $\sqrt{(2/3)}$ makes it orthogonal. By considering the changes implied in the magnitudes of voltages and currents, compared with those from eqns. 3.14 and 3.20, it is readily seen that the corresponding 2-axis model has d and q coils with $\sqrt{(3/2)}$ times the number of turns of the a, b, c coils. It is interesting to note, as occurs in other analyses, that the orthogonal transformation does not correspond to any particularly meaningful physical situation. There are, as Concordia[20] has noted, an infinite number of alternative voltage/current transformation pairs which agree with tensor theory; i.e. which are power-invariant. Their physical interpretation is simply that of an infinite range of possible d, q to a, b, c turns ratios.

An important difference in the treatment of Reference 16, however, is that the per-unit transformation equations are then defined to have an identical form to the ordinary equations (eqns. 3.31 and 3.32), in contrast with the treatment in this Monograph. This may be referred to as the 'new' definition, and it certainly has the merit of simplicity—for example the d, q, z and a, b, c variables share the same voltage–current base product. There are several disadvantages, however, some of which are fundamental, which will be discussed.

First, in much of the literature there is a long-standing preference for per-unit equations of the form of eqns. 3.1 and 3.2, and a number of equations based on this choice (e.g. eqns. 3.23–3.30) are relatively well known. The new definition results in changes to the forms of eqns. 3.23, 3.25 and 3.26, while eqns. 3.28 and 3.30 remain unchanged; the change in the number of turns in the d and q coils compensates for the change in the voltage–current base product. Other changes are implied for some of the power equations, the preferred forms of which are discussed in Section 4.1.

Secondly, the uniqueness of the new definition rests on the choice of the simplest (i.e. orthogonal) algebraic transformation, and it has been noted that this corresponds to a $d, q/a, b, c$ turns ratio of $\sqrt{(3/2)}$. The preferred system eliminates numerical factors arising from a particular choice of turns ratio. Any power-invariant transformation of ordinary voltages and currents would assume the same per-unit form as eqns. 3.1 and 3.2 after normalisation; for each different choice of $d, q/a, b, c$ turns ratio would imply a different choice of the 2-axis voltage and current bases, precisely to achieve this. It will be seen that these arguments centre on the treatment of the d and q variables; the difference in the treatment of the z variables between Reference 16 and this Section is regarded as of lesser importance. The uniqueness of the preferred system lies in the unit-to-unit relationship which it establishes between the d, q and the a, b, c variables, as was noted in eqns. 3.8 and 3.10–3.13.

There are important consequences of this. In the preferred system, the statement for balanced operation that, for example, 'the direct-axis current is 0·5', means either that the 2-axis current i_d is 0·5, or that the phasor component of the per-phase armature current $|\boldsymbol{I}_a|$ is 0·5. Indeed, it means broadly that the armature m.m.f. acting on the direct axis is half of that which would be exerted by the full-load armature currents acting wholly on the direct axis. In the new system, this is not so, and, for example, $|\boldsymbol{I}_a| = 0\cdot408$ when $i_d = 0\cdot5$. Another way of stating this is to say that, in the preferred system, unitary variables are associated with the rated condition; this is so regardless of whether the variables in question are per-phase in the 3-phase coils, or 2-axis variables in the d and q coils.

Thirdly, the underlying principle of the preferred system extends to cover a polyphase armature with any number of phases; i.e. the statement that $i_d = |\boldsymbol{I}_d|$ etc. is always in accordance with eqns. 3.10–3.13. This is briefly demonstrated for the 2-phase case in Section 3.1.3. The principle of the new system, however, gives a different arithmetical relationship between i_d and $|\boldsymbol{I}_d|$ for each different number of phases. It thus attributes to phase number a degree of importance in the per-unit system which is not reflected in the physical behaviour of a polyphase machine. This seems undesirable, since the idea of a per-unit system, is, in effect, to suppress the less-significant features of an analysis.

Since the 3-phase currents are normalised to i_{ao}, it may seem that the need to normalise the d, q, z and rotor currents to a base of $3/2\, i_{ao}$ is a disadvantage of the preferred system. It is in fact of fundamental significance, and demonstrates the principle—already mentioned in Section 2.1, page 19—of 'equal effect'. To exemplify this, compare the effects of the following:

(a) unit current $(3/2)i_{ao}$ acting in coil d
(b) unit current $(3/2)i_{ao}$ acting in a hypothetical field winding constructed identically to coil d
(c) rated 3-phase currents of peak value i_{ao} flowing in coils a, b and c in phase relationships to the rotor such that maximum resultant m.m.f. acts on the direct axis.

From the previous discussion it should be clear that all of these situations produce the same total m.m.f. acting on the direct axis, and this implies that they produce the same amplitude of fundamental air-gap flux-density wave. Thus, in a sense which is important for the operation of the machine, all of the situations (a)–(c) are 'equally effective', and the difference in base currents between (a) and (b) on the one hand, and (c) on the other, is an essential part of this situation.

When considering practical field and other rotor windings which are not constructed identically to coil d, the need arises to define their turns ratios with respect to the stator winding. The aspects of this problem are discussed in Section 3.2, and the notion of equal effect proves to be helpful and important.

A final word may be said about the 'new' definition in relation to the 'X_{ad}' principle for the definition of secondary-current bases. This is a sound and widely used principle (discussed in Section 3.2.1) and is based on an 'equal effect'. Because of this, however, its advantages are largely lost when it is applied in combination with the new definition which is not so based. First, the per-unit 'leakage' inductance $\boldsymbol{L}_{dd} - \boldsymbol{L}_{df}$ of the direct-axis equivalent circuit does not equal the per-unit stator 'leakage' inductance $\boldsymbol{L}_d - \boldsymbol{L}_{ad}$, and indeed it may be negative. Secondly, an unfortunate situation arises in which the unit field current becomes equal in 'effect' to the rated, balanced, 3-phase stator currents, while unit direct-axis current does not. The point in regard to leakage inductance is demonstrated in the footnote,† which assumes a familiarity with the later discussion of Section 3.2.1.

† With the particular transformation of eqns. 3.14 and 3.20, it has already been noted (prior to eqn. 3.26) that
$$L_{df} = L_{afd} \tag{A}$$
The principle of the 'X_{ad}' base always ensures that
$$\boldsymbol{L}_{afd} = \boldsymbol{L}_{ad} \tag{B}$$
and with the preferred system, it follows from eqns. 3.26 that
$$\boldsymbol{L}_{df} = \boldsymbol{L}_{ad} \tag{C}$$
It is already established in eqn. 3.30 that $\boldsymbol{L}_{dd} = \boldsymbol{L}_d$, and it follows that the per-unit direct-axis leakage inductance $\boldsymbol{L}_{dd} - \boldsymbol{L}_{df}$ equals the per-unit armature leakage inductance $\boldsymbol{L}_d - \boldsymbol{L}_{ad}$.

With the new system, the turns in the d coil are increased by a factor $\sqrt{(3/2)}$. Consequently, in contrast to eqn. 3.29,
$$L_{dd} = L_d \tag{D}$$
but both sides of this equation are now to be normalised to the same base and once again eqn. 3.30 is obtained, i.e. $\boldsymbol{L}_{dd} = \boldsymbol{L}_d$. However, because of the increase in the d coil turns, in contrast to eqn. A we have
$$L_{df} = \sqrt{(3/2)}\, L_{afd} \tag{E}$$
and, after the application of a per-unit principle which ensures that eqn. B is satisfied, we have
$$\boldsymbol{L}_{df} = \sqrt{(3/2)}\, \boldsymbol{L}_{ad} \tag{F}$$
Thus the direct-axis leakage inductance $\boldsymbol{L}_{dd} - \boldsymbol{L}_{df}$ is equal to $\boldsymbol{L}_d - \sqrt{(3/2)}\, \boldsymbol{L}_{ad}$, which is not meaningful and could be negative in practice.

The main line of development of this Section has been the establishment of a preferred fundamental 3-phase/2-axis per-unit system, and it will be clear that a definition incorporating eqns. 3.1 and 3.2 is much preferred to any other. It should be said, however, that there are simplicities in the new definition which are attractive, and for a well reasoned account of these and a review of many of the problems of per-unit systems, Reference 16 is recommended.

Figs. 5 and 6 summarise some of the more important results of this Section and of Appendix 6.1. They show the d, q and z equivalent circuits in ordinary and per-unit form, respectively. It must be emphasised that Fig. 5 is based on the assumption of the particular ordinary-transformation equations (eqns. 3.14 and 3.20), and Fig. 6 is based on the preferred form of the per-unit transformation equations (eqns. 3.1 and 3.2). One damper winding is shown in each axis for generality. These have not been included in the analysis here or in Appendix 6.1, but their behaviour is in keeping with that of the field winding, as analysed in the Appendix. Fig. 6 shows only the properties which are an essential consequence of using the per-unit transformation equations (eqns. 3.1 and 3.2); i.e. it does not show any of the particular values appropriate to some circuit elements when particular definitions for rotor-current bases are used. This step is deferred to Section 3.2.

It will be noted that the flux linkages ψ_d and ψ_q appear in the equivalent circuits, both in ordinary and per-unit forms. The general (ordinary or per-unit) equation for the total flux linkage with any coil i of a set of n mutually coupled coils should be noted:

$$\psi_i = \sum_{j=1}^{n} L_{ij} i_j \qquad (3.33)$$

Eqns. 6.6 and 6.8 of Appendix 6.1 are particular examples of this. While in some ways it would be helpful to examine here the manner in which the flux linkages are normalised, to justify their inclusion in Fig. 6, it is on balance better to defer this matter until Section 4.3, where it can be considered together with the closely related matter of normalised time.

In regard to the transformation equations which must relate flux linkages with the d, q and z circuits to those with the a, b and c circuits,

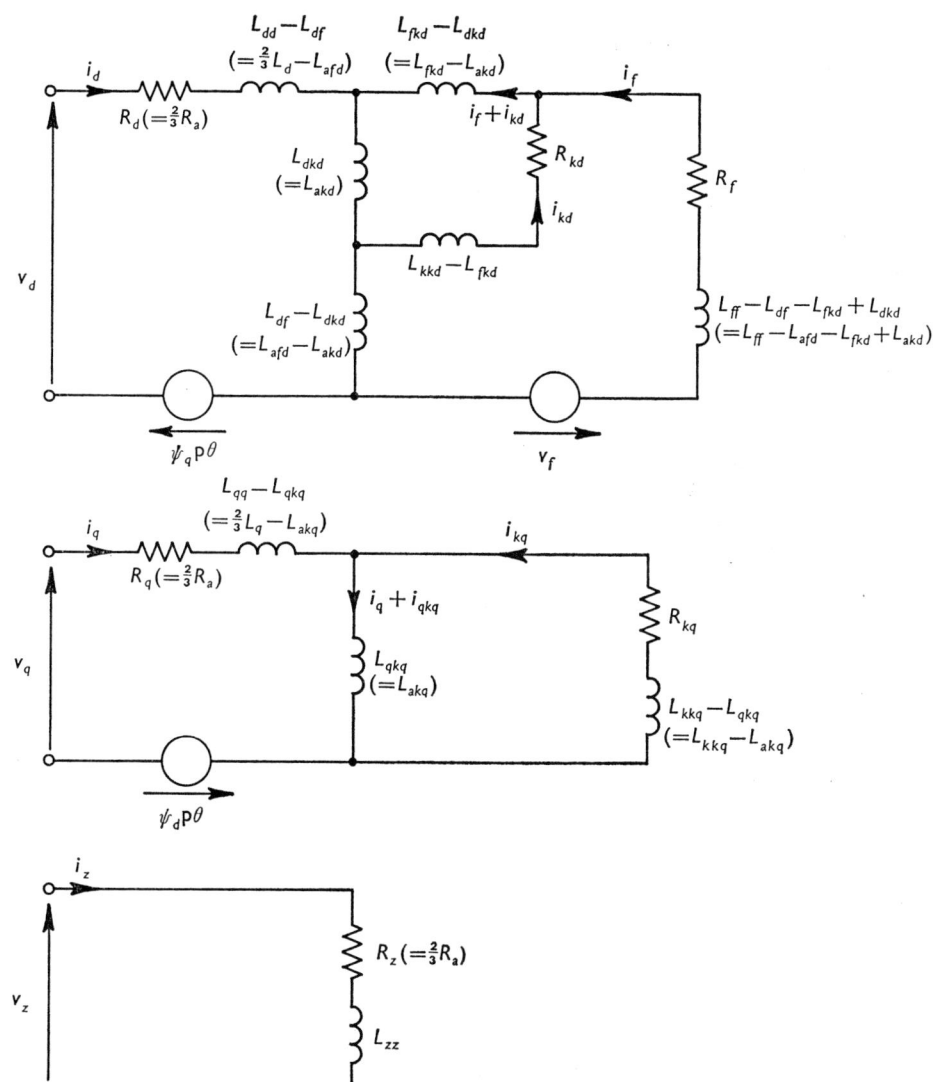

Fig. 5 Equivalent circuits in ordinary parameters

a For direct-axis variables   *b* For quadrature-axis variables   *c* For zero-sequence variables

The values shown in brackets are particular identities which arise from the use of the ordinary transformation equations (eqns. 3.14 and 3.20), corresponding to the d and q and the a, b and c coils having the same number of effective turns. Further identities between the d, q, z and the a, b, c parameters are given in Appendix 6.1. Since no turns ratios have been applied to the secondary circuits, many of the inductances in the d axis and q axis are physically meaningless and could have negative values

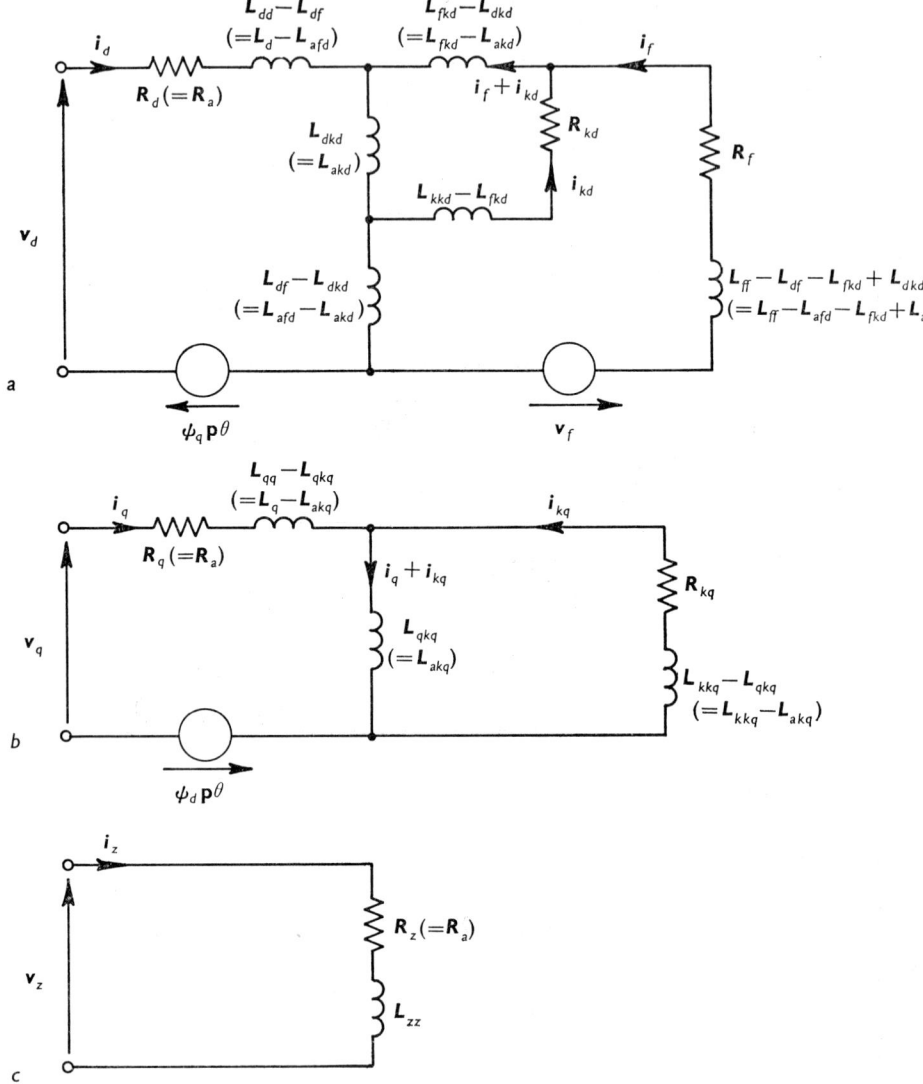

Fig. 6 Equivalent circuits in per-unit parameters

a For direct-axis variables   *b* For quadrature-axis variables   *c* For zero-sequence variables

The values shown in brackets are particular identities which arise from the use of the preferred per unit transformation equations (eqns. 3.1 and 3.2). No identities or simplifications which are dependent on a particular choice of rotor base currents are included here—see Fig. 10. Since no turns ratios have been applied to the secondary circuits, many of the inductances in the d axis and q axis are physically meaningless and could have negative values these may be expected to have in ordinary parameters the same form as the equations for transformation of voltages, rather than of currents. As can be seen from eqn. 3.33, the flux linkages are not related directly to currents, but rather to inductance–current products. By comparison with the form of eqn. 3.14, we have

$$\begin{bmatrix} \psi_d \\ \psi_q \\ \psi_z \end{bmatrix} = \tfrac{2}{3} \begin{bmatrix} \cos\theta & \cos(\theta-120°) & \cos(\theta-240°) \\ -\sin\theta & -\sin(\theta-120°) & -\sin(\theta-240°) \\ \tfrac{1}{2} & \tfrac{1}{2} & \tfrac{1}{2} \end{bmatrix} \begin{bmatrix} \psi_a \\ \psi_b \\ \psi_c \end{bmatrix} \quad (3.34)$$

Eqn. 3.34 may be seen to be consistent with the notion of the 2-axis-machine model. For example, consider the particular condition in which the machine is stationary at $\theta = 0$, a steady current flow in the field winding only producing a flux pattern centred on the vertical axis in Fig. 4(i). By inspection of this Figure, we can write

$$\psi_b = \psi_c = -\psi_a \cos 60° = -\psi_a/2 \quad (3.35)$$

Eqn. 3.34 gives an expression for ψ_d:

$$\psi_d = \tfrac{2}{3}\left(\psi_a - \frac{\psi_b}{2} - \frac{\psi_c}{2}\right)$$

and substituting from eqn. 3.35, we obtain

$$\psi_d = \psi_a \quad (3.36)$$

which is the correct result for the model, remembering that the a and d coils have equal numbers of effective turns.

3.1.2 Principles of preferred per-unit system

The discussion of the previous Section, particularly in regard to the relative merits of the 'new' and 'preferred' 3-phase/2-axis per-unit systems, has brought out several points of fundamental interest. These are worthy of a brief summary, since they may be regarded as useful guiding principles in the establishment of any per-unit system; they will be of considerable help, for example, in later Sections of this Chapter and in Chapter 4. There are three points and these are set out below, with brief reminders in brackets of the ways in which the preferred 3-phase/2-axis per-unit system is consistent with them.

(*a*) In regard to the base values of variables which are open to free choice—i.e. which are independent—the choice should preferably be such that per-unit variables assume unit value, either under the condition of normal rated operation, or under a condition meaningfully associated with it (e.g. in balanced rated operation, peak $\boldsymbol{v}_a$ and peak $\boldsymbol{i}_a$ assume unit value). If balanced, rated stator currents are arranged to act purely on the direct axis—which is not the rated condition but is for the present purpose a condition meaningfully associated with it—then i_d assumes unit value. Comparable arguments apply for i_q, v_d and $\boldsymbol{v}_q$.

(*b*) Broadly consistent with the above principle is the notion that unit current in different windings should preferably be defined so as to produce equal physical effect. Since any current produces several different 'effects' one must proceed to say that the effect in question is (in broad terms) that which is of the most fundamental importance for the intended operation of the machine. (For example, unit i_d produces the same strength of fundamental flux-density wave centred on the direct axis as unit $|\boldsymbol{I}_d|$ in balanced 3-phase operation. Equally, unit i_q produces the same strength of fundamental flux-density wave centred on the quadrature axis as unit $|\boldsymbol{I}_q|$ in balanced 3-phase operation.)

(*c*) In regard to balanced polyphase windings, which in the condition of normal rated operation are broadly intended to act together rather than individually, these should be considered *en bloc* in the per-unit system. The 'effect' is that produced by their simultaneous operation in a normal balanced manner (e.g. unit i_d does not produce an effect equal to that of unit $|\boldsymbol{I}_d|$ acting in one phase coil alone, but rather, as already noted, an effect equal to unit $|\boldsymbol{I}_d|$ in normal balanced polyphase operation.)

The foregoing points are not precise rules by which a per-unit system can be defined in any circumstance. Each particular field of analysis poses its particular problems for the establishment of a preferred per-unit system, and the three points are guiding principles against which a tentative system can be assessed.

3.1.3 2-phase/2-axis transformation

It is interesting to examine briefly the considerations which arise for the 2-axis per-unit analysis of a 2-phase machine, as opposed to those for the 3-phase machine of Section 3.1.1. A suitable transformation between ordinary instantaneous 2-phase (α, β) voltages and d, q voltages† is as follows[7]:

$$\begin{bmatrix} v_d \\ v_q \end{bmatrix} = \begin{bmatrix} \cos\theta & \sin\theta \\ -\sin\theta & \cos\theta \end{bmatrix} \begin{bmatrix} v_\alpha \\ v_\beta \end{bmatrix} \qquad (3.37)$$

where θ is the instantaneous electrical angle between the field winding and the phase α. The inverse of eqn. 3.37 is

$$\begin{bmatrix} v_\alpha \\ v_\beta \end{bmatrix} = \begin{bmatrix} \cos\theta & -\sin\theta \\ \sin\theta & \cos\theta \end{bmatrix} \begin{bmatrix} v_d \\ v_q \end{bmatrix} \qquad (3.38)$$

The transformation of currents which maintains power invariance is, from eqns. 2.26 and 2.29,

$$\begin{bmatrix} i_d \\ i_q \end{bmatrix} = \begin{bmatrix} \cos\theta & \sin\theta \\ -\sin\theta & \cos\theta \end{bmatrix} \begin{bmatrix} i_\alpha \\ i_\beta \end{bmatrix} \qquad (3.39)$$

which has the inverse

$$\begin{bmatrix} i_\alpha \\ i_\beta \end{bmatrix} = \begin{bmatrix} \cos\theta & -\sin\theta \\ \sin\theta & \cos\theta \end{bmatrix} \begin{bmatrix} i_d \\ i_q \end{bmatrix} \qquad (3.40)$$

Clearly, this particular choice of transformations is orthogonal, voltages and currents transforming in an identical manner. Additionally, however, eqns. 3.37–3.40 are those which correspond to the equivalent d and q coils having the same number of effective turns as the α and β coils; this in fact is the primary reason for their choice in the present instance. This may be demonstrated by considering particular conditions of operation in the usual manner by making reference to Fig. 7, which shows the equivalence between the α, β coils and the pseudo-stationary d, q coils in the 2-axis model. Thus, considering the machine to be stationary in the position $\theta = 0$, and considering alternating excitation applied to the field winding, we see from eqn. 3.37 that $v_d = v_\alpha$, which is consistent with the d and α coils having equal effective turns. Comparable simple checks show that the same is true of q and

† The zero-sequence component is absent in the 2-phase machine.

β coils, and any checks will be found to be consistent with this same physical picture. For the 2-phase machine, therefore (unlike the 3-phase one), the orthogonal transformation has the particular physical significance that there are equal turns in the phase and axis coils.

Moreover, although eqns. 3.37–3.40 are in ordinary variables, they have immediately the unit-to-unit properties which were established only with difficulty for the per-unit equations of the 3-phase machine

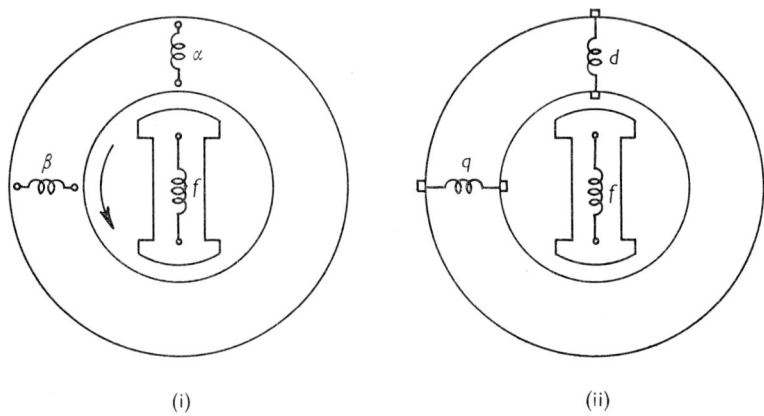

Fig. 7 Coil configurations
(i) The configuration of α and β coils relative to the field winding, with relative motion between the stator and the rotor
(ii) The pseudostationary d and q coils which can replace the stator

in Section 3.1.1; i.e. considering balanced 2-phase operation, we may write

$$\left. \begin{array}{l} v_\alpha = -\hat{v}\sin\omega t \\ v_\beta = -\hat{v}\sin(\omega t - 90°) \end{array} \right\} \quad (3.41)$$

$$\theta = \omega t - \delta \quad (3.42)$$

From eqn. 3.37, it follows that

$$\left. \begin{array}{l} v_d = -\hat{v}\sin\delta \\ v_q = \hat{v}\cos\delta \end{array} \right\} \quad (3.43)$$

whence

$$\hat{v} = \sqrt{(v_d^2 + v_q^2)} \quad (3.44)$$

Eqns. 3.41–3.44 in ordinary variables may be compared with the per-unit equations (3.5–3.8) of Section 3.1.1. The unit-to-unit form of eqn. 3.44 for ordinary voltages can be easily demonstrated for ordinary currents also in the present case, since currents and voltages transform identically.

Thus, it is only necessary to normalise eqns. 3.37 and 3.38 throughout to the single voltage base v_{ao}, and eqns. 3.39 and 3.40 throughout to the single current base i_{ao}, and the equations of transformation for the preferred 2-phase/2-axis per-unit system are obtained. The preferred per-unit transformation equations are identical in form to the orthogonal ordinary equations—in total contrast to the 3-phase machine—and the same voltage–current base product $v_{ao}i_{ao}$ applies in both phase and axis reference frames, again in contrast to the 3-phase machine. As in Section 3.1.1, v_{ao} and i_{ao} are taken as the peak rated values per phase.

The analysis of the 2-phase situation is, for the present purpose, completed by noting some results (now using per-unit phasor notation) which are equivalent to eqns. 3.10–3.13—again for balanced operation. These results are shown to follow from eqns. 3.37–3.43. Thus

$$\boldsymbol{V}_a = \boldsymbol{V}_d + \boldsymbol{V}_q \qquad (3.45)$$

where

$$\left.\begin{array}{l}|\boldsymbol{V}_d| = v_d \\ |\boldsymbol{V}_q| = v_q\end{array}\right\} \qquad (3.46)$$

and correspondingly for currents,

$$\boldsymbol{I}_a = \boldsymbol{I}_d + \boldsymbol{I}_q \qquad (3.47)$$

where

$$\left.\begin{array}{l}|\boldsymbol{I}_d| = i_d \\ |\boldsymbol{I}_q| = i_q\end{array}\right\} \qquad (3.48)$$

It will be recalled that the difference between the voltage–current base products in the phase and axis circuits of the 3-phase machine was introduced, basically, to obtain equations of the form of eqns. 3.45–3.48. In the 2-phase machine the need for this difference does not arise. The equations are simply obtained, and the associated benefits for the per-unit system include conformity to the principles of Section 3.1.2. Thus, in per-unit terms, the 2-phase machine may be regarded as the fundamental polyphase machine; the per-unit analysis of the 3-phase machine is rendered similar to that of the 2-phase machine by the use of varied voltage–current base products.

3.1.4 Symmetrical-components transformation

Since the analysis of Section 3.1.1 involved some consideration of the zero-sequence variables, it is appropriate to consider briefly the family of 3-phase symmetrical-component variables which includes the zero-sequence variable. This is the more worthwhile, since the subject provides an illustration of two points which were made in earlier Sections; first, as noted in Section 2.2, it is not always possible to construct simply a new physical model corresponding to the power-invariant transformation; secondly, as noted in Section 3.1.2, the problem of establishing a per-unit system in a particular field of analysis has to be considered on its merits, and no precise rules apply.

The zero-sequence variables in Section 3.1.1 were defined for instantaneous parameters. It is possible to define positive- and negative-sequence variables for instantaneous parameters also, but we confine ourselves to the main area of application of the group of symmetrical components; namely to steady-state, unbalanced operation, analysed in terms of complex phasors. As is well known (see, for example, Reference 11, which is recommended for a brief background theory of symmetrical components), there are two commonly quoted forms of the symmetrical-component/3-phase transformation equations in ordinary parameters. The first form is that corresponding to power-invariance and orthogonality; i.e.

$$\begin{bmatrix} V_z \\ V_p \\ V_n \end{bmatrix} = \sqrt{\tfrac{1}{3}} \begin{bmatrix} 1 & 1 & 1 \\ 1 & a & a^2 \\ 1 & a^2 & a \end{bmatrix} \begin{bmatrix} V_a \\ V_b \\ V_c \end{bmatrix} \qquad (3.49)$$

which has the inverse

$$\begin{bmatrix} V_a \\ V_b \\ V_c \end{bmatrix} = \sqrt{\tfrac{1}{3}} \begin{bmatrix} 1 & 1 & 1 \\ 1 & a^2 & a \\ 1 & a & a^2 \end{bmatrix} \begin{bmatrix} V_z \\ V_p \\ V_n \end{bmatrix} \qquad (3.50)$$

The form of the current transformations is of course identical to that of eqns. 3.49 and 3.50.

Alternatively, there are the transformation equations commonly used in power-systems analysis, which differ from the above by a factor of $1/\sqrt{3}$; namely

$$\begin{bmatrix} V_z \\ V_p \\ V_n \end{bmatrix} = \tfrac{1}{3} \begin{bmatrix} 1 & 1 & 1 \\ 1 & a & a^2 \\ 1 & a^2 & a \end{bmatrix} \begin{bmatrix} V_a \\ V_b \\ V_c \end{bmatrix} \tag{3.51}$$

with the inverse

$$\begin{bmatrix} V_a \\ V_b \\ V_c \end{bmatrix} = \begin{bmatrix} 1 & 1 & 1 \\ 1 & a^2 & a \\ 1 & a & a^2 \end{bmatrix} \begin{bmatrix} V_z \\ V_p \\ V_n \end{bmatrix} \tag{3.52}$$

The form of the current-transformation equations is again defined to be identical in form to the voltage-transformation equations (eqns. 3.51 and 3.52). This combination of transformations is clearly not power-invariant in the tensor sense.

Considering more fully the power-invariant formulation of eqns. 3.49 and 3.50, it is reasonable to try to construct a physical model in the (z, p, n) reference frame. In Section 3.1.1, for instance, the notions of 'the d coil' and 'the q coil' were established for the 3-phase/2-axis transformation. These coils (which were pseudostationary, i.e. commutated with motional voltages induced in them) when excited by v_d, i_d and v_q, i_q, respectively, constituted complete electromagnetic replacements for the a, b and c coils, given that the zero-sequence component of stator excitation was zero. We were primarily concerned with defining a per-unit system with reference to balanced operation, and accordingly there was no pressing need to discuss the physical nature of 'the z coil'. For the 2-phase machine, the matter cannot arise, since zero-sequence variables are automatically absent. In this Section, the sequence components, not only for the z variables, but also for the p and n variables, are of central interest, and it will be seen that a simple physical interpretation in terms of new p, n and z coils is not available.

It is well known[11] that the mathematical behaviour of the positive-sequence variables, for example, is consistent with their producing a pure forward-rotating heteropolar field in the air gap of the machine.

A first step in establishing a simple model is, therefore, the identification of a single 'p coil' which excites an identical field. Since this 'coil' must be stationary with respect to the stator (the frequency of the a, b, c and z, p, n variables being the same), and since a pure rotating field cannot be produced by any excitation of a single stationary coil, it is immediately clear that no 'p coil', and hence no physical interpretation of the symmetrical component variable analogous to that of the 2-axis variables, is available. A model involving a single rotating p coil is not permissible as a further transformation would be implied. This situation and its consequences are consistent with the usual opinion that the z, p and n variables represent components of the a, b and c variables in the original a, b and c coils. However, the magnitudes of the symmetrical-component voltages applied to each phase coil are not equal to the magnitudes of the defined symmetrical-component voltage variables; they differ by a factor $1/\sqrt{3}$. This is also true for currents, and these relationships are illustrated in Fig. 8.

Turning now to the choice of a per-unit system for the symmetrical-component equations, and considering first the power-invariant transformation equations (eqns. 3.49 and 3.50), an obvious procedure would be to normalise throughout the ordinary voltage variables to a base v_{ao}. In the identical current-transformation equations, the ordinary current variables would be normalised to a base i_{ao}. The form of the per-unit transformation equations would thus be identical with that of the ordinary orthogonal equations —a result comparable with that for the 'new' per-unit 3-phase/2-axis system, discussed in Section 3.1.1. As with that system, the present procedure is mathematically simple, but it leads to some results for the per-unit system which are not desirable.† For example, consider that the 3-phase supply to the machine has a balanced rated value. The per-unit phase voltages can then be written as

$$\left. \begin{array}{l} V_a = e^{j\omega t} \\ V_b = a^2 e^{j\omega t} \\ V_c = a e^{j\omega t} \end{array} \right\} \tag{3.53}$$

† It may be noted for general interest that the instantaneous zero-sequence component of the 'new' definition corresponds to eqn. 3.49, whereas that of the established and 'preferred' definition corresponds to eqn. 3.51.

whence, with a per-unit transformation equation of the form of eqn. 3.49, it follows:

$$V_z = 0$$
$$V_p = \sqrt{3}\, e^{j\omega t}$$
$$V_n = 0$$
(3.54)

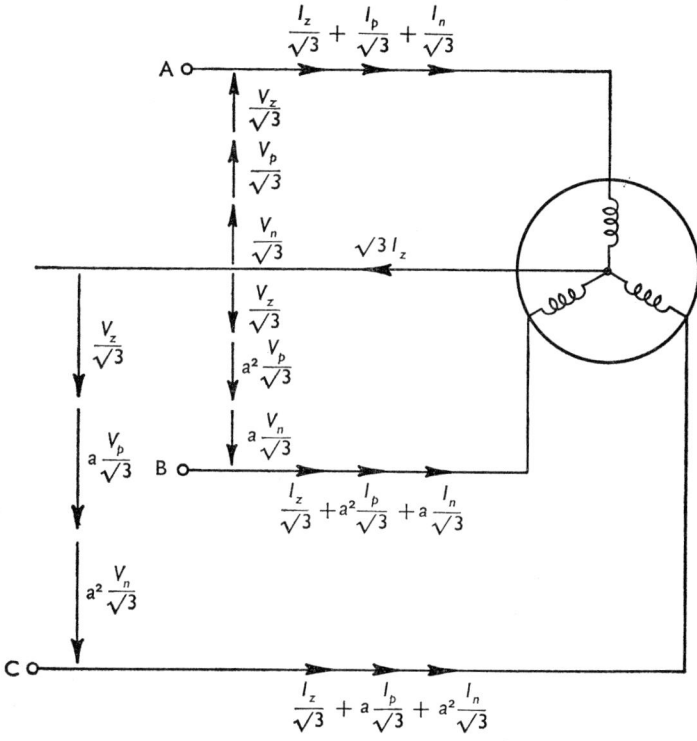

Fig. 8   General 3-phase machine, showing stator excitation in terms of power-invariant symmetrical-component ordinary variables

Thus, we have a system in which the positive-sequence voltage amplitude is $\sqrt{3}$ when the terminal phase voltages of the machine contain only a positive-sequence component and the phase voltage amplitude is unity. Similar arguments apply for negative- and zero-sequence voltages and currents. To express this differently, operation under conditions of rated excitation corresponds to positive-sequence voltage and current of $\sqrt{3}$; whereas unity would be expected, and would agree much better with (a) in Section 3.1.2.

Conversely, the 'power-systems' equations (eqns. 3.51 and 3.52) may be normalised throughout to a base v_{ao}, and the identical current-transformation equations may be normalised to a base i_{ao}. In balanced rated operation, represented by the per-unit equations (eqns. 3.53), it follows from eqn. 3.51 that

$$\left. \begin{aligned} V_z &= 0 \\ V_p &= e^{j\omega t} \\ V_n &= 0 \end{aligned} \right\} \quad (3.55)$$

Thus this system makes rated excitation correspond to a positive-sequence voltage and current of unity. This now agrees with (a) in Section 3.1.2, but it is achieved at the expense of the loss of power-invariance in the transformation.

For reasons given below, this loss of power invariance is not serious for the majority of applications of symmetrical-component theory; this explains the established popularity of eqns. 3.51 and 3.52 in the literature for ordinary and per-unit working. One result of this non-power-invariant transformation is that the mutual inductances with any other coupled circuit x become asymmetric in the (z, p, n, x) reference frame. This would be serious enough, as we have seen in Sections 2.1 and 2.2, but the normal field of application for symmetrical components is in problems in which other circuits are not represented directly, but are represented as effects referred to the a, b and c circuits so that no problems arise. We can therefore say that the form of eqn. 3.51 would be preferred for per-unit voltages and currents in normal applications.

Finally, it is of some interest to draw a parallel between the treatment which was accorded to the 3-phase/2-axis transformation in Section 3.1.1, and that which could be applied to the present case. It is quite possible to start with the power-invariant form of eqns. 3.49 and 3.50 (and identical forms for currents), and to normalise the a, b, c voltages to a base v_{ao} and the z, p, n voltages to a base $\sqrt{3}v_{ao}$. A similar procedure may be followed for currents. This results in voltage–current base products $v_{ao} i_{ao}$ in the (a, b, c) reference frame and $3v_{ao}i_{ao}$ in the (z, p, n) reference frame. The per-unit transformation equations then assume the preferred form of eqns. 3.51 and 3.52 for per-unit voltages and currents. The logical justification for this step is that, in both reference frames, the base voltages and currents have now been chosen as the values corresponding to rated operation. The benefit is a unit–unit relationship between the per-unit variables in the two reference frames.

3.2 Secondary-current bases of the synchronous machine
3.2.1 X_{ad} base

With the X_{ad} base, the rotor current in any circuit is taken to be that current which induces in each phase a voltage $X_{ad}i_{ao}$ in a balanced synchronous operation. This is the voltage caused by the armature reaction on the direct axis, and clearly the principle of equal effect [expressed in (b) in Section 3.1.2] is met by this definition; the rotor current induces the same voltage, and hence *the same air-gap flux wave, as balanced 3-phase unit-peak armature currents* (acting on the direct axis). The fluxes produced in the air gap of a machine are central to all aspects of machine performance, and physically the only meaningful way in which one winding (or set of windings) can be 'evaluated' in terms of another is through its effectiveness in developing the basic flux wave. In view of this, it is not surprising that the X_{ad} base leads to a per-unit system which reflects most closely the physical features of the machine and that, consequently, it is 'well behaved' in a number of ways that will become apparent.

The points concerning the flux wave in the air gap—its use as a measure of winding effectiveness and its basic importance for machine performance—apply strictly to the actual distribution of flux, and this in general is nonsinusoidal. It is, however, a standard practice to base all discussion of the X_{ad} base on the assumption of purely sinusoidal fluxes—hence the quantity $X_{ad}i_{ao}$, which implies only fundamental quantities. This assumption is completely justified, since it is basic to the whole of the general body of synchronous-machine theory whether it is developed in terms of phasor diagrams or formally in terms of the unified theory relying on transformations.

For normal operation of a synchronous machine with a d.c.-excited field winding and the stator excited by balanced polyphase currents, the complex pattern of the flux density in the air gap can be resolved into fundamental plus harmonic travelling waves. The fundamental of course is predominant, and is the only one which both induces a fundamental-frequency voltage in the stator winding and rotates synchronously with the rotor; i.e. it interacts with both windings in the basic synchronous manner. The coupling between the windings caused by the harmonic fluxes is a small proportion of the total mutual coupling and does not contribute to the fundamental synchronous operation, although small power flows may be associated with the harmonics. Therefore, in the basic analysis, it is sensible to ignore it by treating the harmonic fluxes in terms of pure leakage-reactance or stray-loss components separately, and expressing these in per-unit terms for the per-unit system based on the ideal machine.

Considering now the particular matter of the interaction between the polyphase stator winding and a full-pitched rotor winding (which includes the field winding to a good approximation), it follows, from the preceding discussion, that it is appropriate to regard these windings as being mutually coupled by the fundamental sine-wave component of the air-gap flux density only. It is thus possible to visualise the flux-coupling situation as ideal—the fundamental air-gap flux providing an ideal pattern of mutual-flux linkage between the stator and field windings. Considering the 2-axis machine model, an ideal turns ratio can be defined, through eqn. 3.23, by selecting an appropriate value of i_{fo}. This value is that which excites the same magnitude of fundamental air-gap flux density as $(\frac{3}{2})i_{ao}$ flowing in the d coil; i.e. as balanced, unit-peak, 3-phase currents flowing in the stator winding and acting on the d axis.

In the practical calculation of i_{fo}, account has to be taken of all the factors influencing the form of the flux-density wave; this means not only the physical disposition of the windings (whether they are concentrated, distributed or short-pitched) but also the variation in the air-gap 'permeance', allowing for simple changes in gap length and more complex exciting-current boundary-shape interactions.† These

---

† In the discussion, as is normal, the effects of saturation are disregarded. Strictly, the flux coupling and hence the turns ratio depend on the particular

factors influence the stator and rotor windings differently in regard to their production of fundamental flux exemplified for the direct-axis position in Fig. 9. In Fig. 9a, the main features of the flux paths are

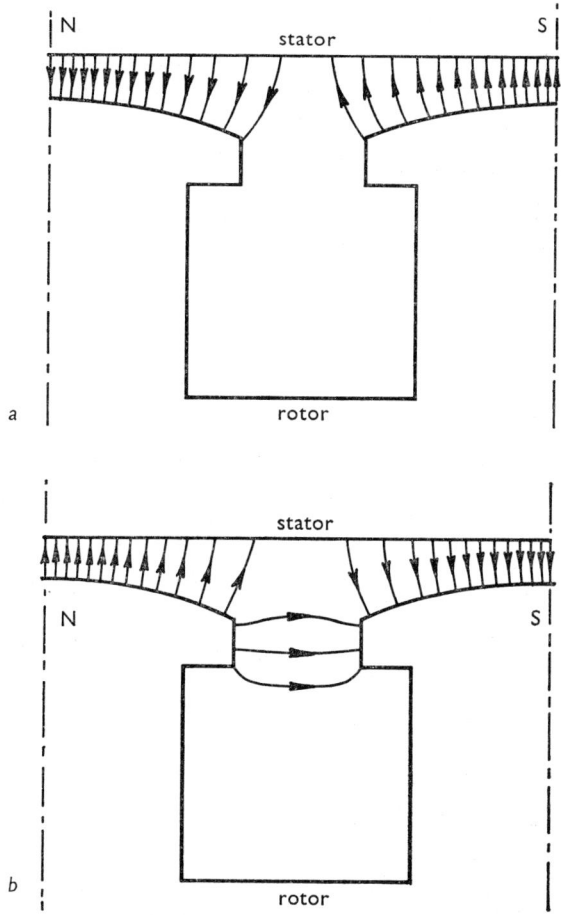

Fig. 9 Air-gap flux distributions produced in salient-pole machine
 a Sinusoidal stator m.m.f. acting on the direct axis
 b Excitation of a concentrated field winding (not shown)

operating condition[26]. It is preferable, however, that an unambiguous definition for turns ratio is provided, and that saturation should subsequently be allowed for by judicious adjustment of the reactances.

indicated for excitation by a sinusoidal direct-axis m.m.f. from the smooth armature, while in Fig. 9*b* the flux distribution is caused by the excitation from the concentrated field winding, the iron being unchanged. An obvious difference in the flux distribution near the pole tips is visible since interpolar leakage occurs in Fig. 9*b* but not in Fig. 9*a*; another more significant difference arises under the pole face, since for Fig. 9*a* there is a sinusoidal variation in the magnetic potential difference across the gap, while for Fig. 9*b* the potential difference is constant. The methods for the determination of the flux distribution in such situations are well known; e.g. by the use of Wieseman's curves[23], Rankin's integral expressions[24], or by analogue or digital computation.

By incorporating the knowledge of the actual flux distribution, the turns ratio of the X_{ad} base for the full-pitched direct-axis rotor windings, including the field, may thus be written as

$$\frac{i_{fo}}{(3/2)i_{ao}} = \frac{4}{\pi} \frac{A_{d1}}{F_{d1}} \frac{N_{aa}}{K_p K_d n N_{ff}} \quad (3.56)$$

Considering the circumstances relating to the quadrature-axis position, it will be apparent that a rigorous application of the equal-effect principle leads to a full-pitched rotor-circuit current base and turns ratio different from the above direct-axis value. This is because the m.m.f./permeance interaction in the quadrature-axis position is again different for excitation from the armature and field sides, and, in general, this difference is not simply related to that occurring for the direct-axis position. However, the rigorously derived ratios for the two axes are not likely to differ greatly, and it is common both in practice and in the general literature (e.g. Rankin[3, 24], Concordia[10], Adkins[8]) to use the same turns ratio for the quadrature-axis as for the direct-axis. One significant advantage of this is that per-unit currents in coincident direct- and quadrature-axis damper circuits can be added directly.

However, it is important to emphasise that, while a single turns ratio can be adopted in this way, this ratio is not ideal as far as the quadrature axis is concerned. Consequently, the components of the equivalent circuit for the quadrature axis must be calculated in accordance with the general method of eqns. 2.38. It is not correct to simply reproportion the secondary leakage inductance by multiplying by (chosen turns ratio/ideal turns ratio)², as this assumes that eqns. 2.41 apply. It is to be emphasised that this point applies whenever nonideal turns ratios are employed.

The direct-axis turns ratio is, as noted, ideal, and, in accordance with the considerations of Section 2.3 leading to eqns. 2.41, the per-unit series inductances of the direct-axis equivalent circuit do correspond with the per-unit leakage inductances calculated by the designer. It should be noted here that, if time is normalised in the preferred manner indicated in Section 4.3, the per-unit inductance becomes equal to the per-unit rated-frequency reactance, and the words 'inductance' and 'reactance' could be interchanged in this discussion, as elsewhere[2,3]. The point about leakage inductance requires some elaboration. The X_{ad} base essentially ensures that unit field current produces the same air-gap (mutual) peak flux linkage with a phase coil as the balanced unit-peak armature currents; i.e. it ensures that

$$L_{ad} = L_{afd} \qquad (3.57)$$

remembering that L_{afd} is evaluated 'one-way' in the manner of eqn. 3.24, Section 3.1.1. Consequently, used in conjunction with the preferred per-unit transformation equations (eqns. 3.1 and 3.2) and not otherwise—see footnote on page 49—which ensures that

$$\left. \begin{array}{l} L_{df} = L_{afd} \\ L_{dd} = L_d \end{array} \right\} \quad \text{(see eqns. 3.26 and 3.30)}$$

the result is obtained:

$$L_{dd} - L_{df} = L_d - L_{ad} \qquad (3.58)$$

The left-hand side of eqn. 3.58 is the per-unit series inductance of the d coil in the direct-axis equivalent circuit, and the right-hand side is the per-unit leakage inductance of the armature as normally calculated by the designer; this is often assumed to be the same on both axes.

The principle of the X_{ad} base can be extended to any number of damper windings on the direct-axis, whether full- or short-pitched, and in each case

$$L_{ad} = L_{afd} = L_{akd} = L_{df} = L_{dkd} \qquad (3.59)$$

in conjunction with the per-unit transformation equations (eqns. 3.1 and 3.2). However, with respect to short-pitched damper windings, the consequence is the appearance of 'unattractive' values, possibly including negative ones, among the various inductances of the rotor windings. (See Section 3.2.2.)

The equivalent circuits for the d and q axes and for the zero-sequence variables are shown in Fig. 10. These follow from Fig. 6 by the use of the X_{ad} base turns ratio and by the representation of one full-pitched damper winding in each axis. Note that, in accordance with eqn. 3.57, the use of the X_{ad} turns ratio permits the replacement of $\boldsymbol{L}_{afd}$ by $\boldsymbol{L}_{ad}$ in the d axis circuit and this is computationally convenient. It is to be realised, however, that the frequently taken step of replacing $\boldsymbol{L}_{afd}$ by $\boldsymbol{L}_{ad}$ in the equivalent circuit for the equal-mutuals base (Section 3.2.4) is strictly incorrect, since eqn. 3.57 holds only for the X_{ad} turns ratio. To emphasise that the X_{ad} turns ratio is not ideal for the quadrature axis, two inequalities are indicated in the q axis circuit.

A simplification of the d axis circuit which is not uncommon is the neglect of the series inducance $\boldsymbol{L}_{fkd} - \boldsymbol{L}_{ad}$. This has been justified on the grounds that, in practice, the flux linking the damper circuit is very nearly equal to that linking the armature. It has also been argued that this condition holds, provided the damper circuit is 'near the air gap', but this is clearly not enough. The ideal circumstance which makes $\boldsymbol{L}_{fkd}$ equal to $\boldsymbol{L}_{ad}$ is that the field, armature and damper circuits all link a single 'ideal' mutual flux, and hence it is evident that the damper circuit must also be effectively full-pitched. This condition is clearly met neither by short-pitched damper circuits nor by solid-pole circuits. In view of this, and the fact that with modern computing methods it is not a significantly helpful simplification, the authors do not recommend its use. In this connection, see also the discussion of the equal-mutuals base in Section 3.2.4.

The following authors are among those who use the X_{ad} base: Adkins[8], Concordia[10], Kilgore[25] and Rankin[2,3,24].

3.2.2 Extension of X_{ad} base to short-pitched damper windings

The principle of the X_{ad} base as discussed above could be applied directly to a damper circuit of less than full pitch. For example, this was indicated by Rankin in one of his papers[3] which led to an equation which is the direct equivalent of eqn. 3.56 for turns ratio. Rankin,

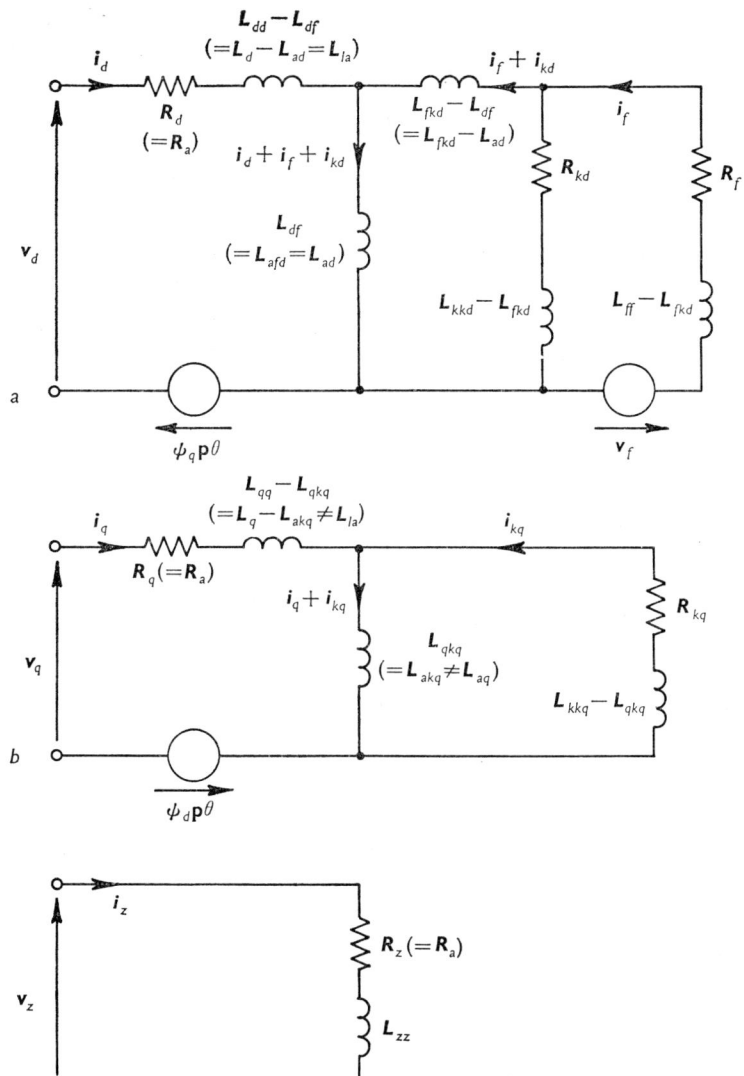

Fig. 10   Equivalent circuits in per-unit parameters

a For direct-axis variables   *b* For quadrature-axis variables
c For zero-sequence variables

The values shown in brackets are particular identities which arise from the use of the preferred per-unit transformation equations (eqns. 3.1 and 3.2), together with the X_{ad} base, for the definition of base currents in the field winding and damper windings, which are assumed to be full-pitched on both axes however, without giving reasons, advised against the use of this procedure and his advice is supported here, for the important reason that the resulting turns ratios are not ideal. In Section 3.2.1, full-pitched concentrated windings, or their equivalent, were involved, permitting complete ideal coupling with the basic fundamental sine wave of flux. However, with the short-pitched coil, since this does not completely embrace the halfwave of flux density, ideal coupling between the damper and armature simply does not exist. It is possible, however, to recognise and use instead a form of ideal coupling between short-pitched dampers on the one hand and a full-pitched damper on the other.

Fig. 11 shows diagrammatically, in developed form, concentrically nested damper circuits on the direct axis of a salient-pole machine, and also the flux density produced in the air gap by currents in the inner circuits 1, the currents being (as in normal operation) of equal magnitude but opposite signs on adjacent poles. As far as interaction with the armature is concerned, this flux distribution emphasises the absence of the ideal coupling just noted, but more important, it can be seen that the coupling of the flux due to damper 1 is complete with all the outer damper circuits 2, 3 and 4. Further, if any one of the outer circuits is excited by a current of the same magnitude as in circuit 1, it can be seen that the flux wave across the span of circuit 1 is identical in magnitude and form to that shown.† Consequently, the flux contained in the hatched area represents an ideal mutual flux, and there is an ideal turns ratio of 1:1 between circuit 1 and all the outer circuits. Similarly, there is an ideal ratio of 1:1 between any circuit and those circuits embracing it.

Thus it is seen that, instead of referring damper circuits directly to the armature (inevitably introducing nonideal turns ratios), it is possible to work instead in terms of two ideal ratios: the first refers all

† This assumes of course that the iron is infinitely permeable, and hence provides equipotential surfaces at the air gap within each current-carrying circuit. In fact, the existence of finite permeability will prevent the surfaces from being truly equipotential[26]. Further, it should be recognised that the saturation effects will generally result in nonreciprocality of mutual inductances between dampers[26]. However, as discussed earlier, it is best to assume that the iron is infinitely permeable in developing basic results, and then to allow later for the influences of finite permeability.

damper circuits on a 1:1 basis to the full-pitched damper circuit (whether this actually exists or not), and the second refers the full-pitched damper circuit to the armature by the ideal ratio discussed above. In effect, this means using a single turns ratio for all the damper

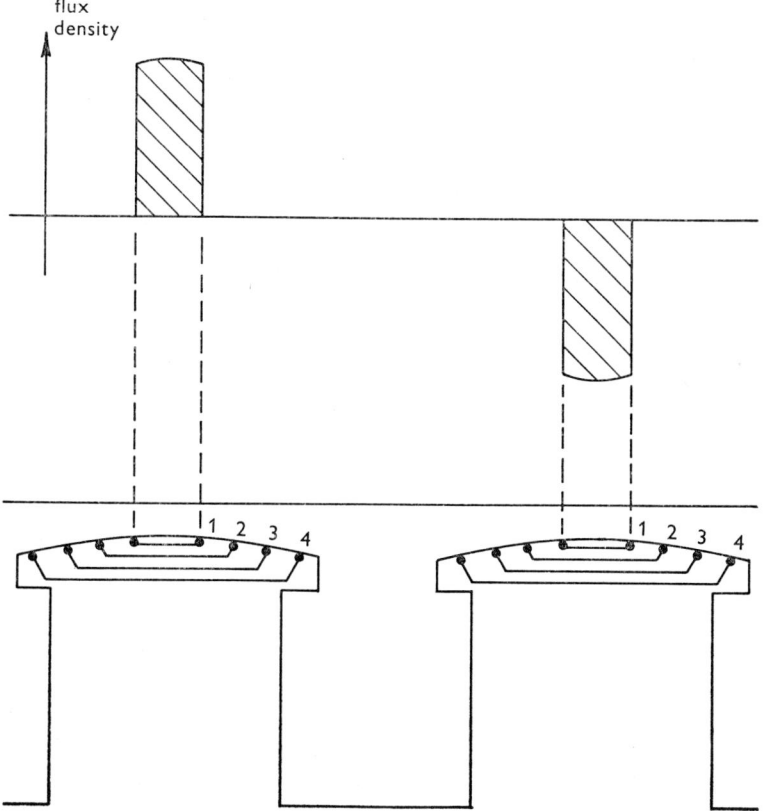

Fig. 11 Concentric direct-axis damper windings, showing pattern of flux density caused by excitation of the inner winding 1 only

circuits, the value of which is given by writing $i_{k0} = i_{f0}$ and $N_{ff} = 1$ in eqn. 3.56; i.e.

$$\frac{i_{k0}}{(3/2)i_{ao}} = \frac{4}{\pi}\frac{A_{d1}}{F_{d1}}\frac{N_{aa}}{K_p K_d n} \qquad (3.60)$$

This ratio corresponds to what may be called 'the full-pitch X_{ad} base'.

The authors strongly recommend its universal adoption for multiple damper-circuit calculations since it is physically the most meaningful,† and, overall, it is computationally the simplest. In effect, it was used by Rankin in a later paper[24]. The complete equivalent circuit for this base, including separate elements for the different effects of gap fluxes, slot- and end-leakage fluxes, and including self and mutual resistance, is given in Appendix 6.3.

3.2.3 M.M.F. base

For the m.m.f. base, the rotor-circuit current base is chosen so that the rotor-circuit m.m.f. per pole is equal to the 'flat-topped armature reaction' of balanced, 3-phase armature currents of unit peak value. The 'flat-topped armature reaction' is an artificial concept, being 3/2 times the m.m.f. due to peak phase current flowing in a notional full-pitched, concentrated winding with turns which produce the same fundamental component of m.m.f. as the actual distributed (and generally non-full-pitched) stator phase winding. It is defined by eqn. 3.56 with A_{d1} and F_{d1} omitted. This m.m.f. wave is clearly rectangular, and the principle of the m.m.f. base may be expressed alternatively as giving unit currents in the rotor and armature windings which produce rectangular m.m.f. waveforms (real for the rotor and hypothetical for the stator) of the same amplitude, but not the same 'widths', since for a damper circuit the latter is obviously dependent on the circuit pitch.

There is no physical significance in this base, and its only merit is the very slight and superficial one that it affords a marginally simpler calculation of the turns ratio than the X_{ad} base. This arises from the fact that calculation of the turns ratio is independent of the flux distribution, being based purely on m.m.f., and ignoring all consideration of the air-gap form. By the same token, however, it generally loses greatly in physical significance, and computations based on it are not likely to reflect the ordinary characteristics of the machine. For example, it has been found by the authors that, with the m.m.f. base, using representative data for a large salient-pole machine, it is possible for

† It avoids, for example, the appearance of negative per-unit impedances in the equivalent circuit, which could arise if the more general X_{ad} base were employed.

the series inductance $L_{ff} - L_{df}$ in the direct axis (neglecting the damper winding) to be negative. It will of course be realised that, for cylindrical air-gap machines, the m.m.f. base yields the same numerical results as the X_{ad} base.

The base has not been used very much, but it was employed by Linville[29] in his well known treatment of multiple damper circuits. In this application, the base automatically results in a 1:1 turns ratio between the damper circuits themselves, and this, as discussed in the previous Section, is recognisable as an ideal ratio. In this particular respect, therefore, the principle of the base is physically meaningful, but this is in itself of little significance since, by the same principle, no ideal ratio can be identified for windings on the armature.

In the treatment of multiple damper circuits (by whatever base system), careful consideration must be given to the flux patterns excited by the short-pitched coils to determine their mutual inductances with the armature windings. Linville's paper may be found helpful in consideration of this matter.

3.2.4 Equal-mutuals base

With this base, as the name implies, the turns ratios are chosen to make equal all the per-unit mutual inductances on an axis. It must be emphasised, however, that the base can be applied, in general, only to the case of three coupled circuits; i.e. to a rotor with a field winding and one damper winding, or with only two damper windings.

The reasons for the restriction to three circuits are as follows. For N arbitrary circuits all mutually coupled, it may be seen that the number of independent mutual reactances is $(N/2)(N-1)$, and that the number of adjustable turns ratios is $N-1$. Therefore, the number of circuits for which the turns ratios can be chosen to make all mutual reactances equal to one free value is obtained by equating the degrees of freedom; i.e.

$$\frac{N}{2}(N-1) - 1 = N - 1 \qquad (3.61)$$

whence $N = 3$. Considering next the particular problem of M nested damper circuits, it is clear that each circuit is characterised by a single mutual reactance with all circuits of greater pitch (see Section 3.2.2), leading to $M-1$ different mutual reactances between the dampers. Additionally, each circuit has a different mutual reactance with the stator direct-axis winding (based only on coupling through the fundamental sine-wave component of the flux density), and the total of these mutuals is M. The number of adjustable turns ratios is equal to the number of damper circuits M; so the permissible value of M for equal-mutual reactances is given by equating the total number of mutual reactances minus one to the adjustable turns ratios; i.e.

$$(2M-1)-1 = M \tag{3.62}$$

whence
$$M = 2$$

Thus again, the total number of possible circuits is three. The only exception to this result which is worth noting is the somewhat unlikely situation of a machine with a uniform air gap and more than three distributed polyphase windings.

Consequently, in applying an equal-mutuals base to an indefinite number of circuits, White and Woodson[7] automatically incur, though it is unlikely that this will be appreciated, an essential restriction on the nature of the machine circuits. Since this restriction is not met by any normal machines with more than three circuits, the discussion is of no real practical significance. Because of the limitation of the base to three circuits, its value for modern computation is limited, and the authors do not recommend it.

The merit of the base is that, for three circuits, it leads to a slightly simpler form of equivalent circuit than that for the X_{ad} base (Fig. 10); the simplification is that, since all the mutual inductances are equal, the series inductance $\boldsymbol{L}_{fkd} - \boldsymbol{L}_{df}$ is zero. It is to be noted, however, that the inductance $\boldsymbol{L}_{df} (= \boldsymbol{L}_{afd})$ cannot be replaced by $\boldsymbol{L}_{ad}$ in the equal-mutuals equivalent circuit, as is possible for the X_{ad} base. This fact has been overlooked by a number of writers.

3.2.5 Unit-voltage base

In this case, the rotor-current base is defined as that which would induce rated voltage in the open-circuited armature. It was designed essentially to deal with the field winding, and it has the superficial advantage of making the stator/rotor mutual coefficients (e.g. $\boldsymbol{L}_{df}$ in circuit terms) equal to unity for the field circuit (or any additional rotor circuit). For example, it has been quite widely used by Doherty and Nickle[27], Park[28], and Stephen[5]. For Park's particular usage, however, Rankin notes[3] that nonreciprocal mutual reactances arise, and this effectively excludes treatment of additional rotor circuits. More generally, as discussed below, the unit-voltage base has serious shortcomings without offering any real advantages, and the authors do not recommend its use.

First, it is distinguished from the other three bases by the fact that the resulting turns ratio can differ substantially from the ideal. For a machine having a per-unit synchronous reactance of the order of unity, the unit-voltage base gives a turns ratio for the field winding approximately equal to that of the X_{ad}, m.m.f. or equal-mutuals bases. But, with a synchronous reactance of about two, which could be achieved by halving the air gap and leaving the windings unchanged, the unit-voltage turns ratio is approximately double the X_{ad} value. This situation arises because unit current is not defined (as with the X_{ad} or m.m.f. bases) to have the same effect (in some sense) as unit current in other windings. Although superficially it appears to be a useful definition, it is not physically meaningful in interrelating different circuits. One consequence of this is that the base is not convenient for use with equivalent circuits; in general, these would contain inductance components totally unrelated to the values normally calculated in design practice, some probably being negative in sign.

Another aspect of the lack of physical meaning of the unit-voltage base is as follows. Consider the hypothetical case of a synchronous induction motor, in which the field winding is mechanically identical to the stator 3-phase winding; i.e. having the same layout and number of turns. The stator carries balanced 3-phase currents, and the rotor is d.c.-excited and is connected as shown in Fig. 12. It should be clear from the discussion in Sections 3.2.1 and 3.2.3 that the X_{ad} and m.m.f. bases would both give the same turns ratio, which from eqn. 3.23 is 2/3. Noting that the ordinary resistance of the 'field' in Fig. 12 is $3/2\ R_a$, the per-unit field resistance can be calculated from eqn. 3.25 as

$$R_f = \left(\frac{2}{3}\right)^2 \left(\frac{3i_{ao}}{2v_{ao}}\right)\left(\frac{3R_a}{2}\right) = \left(\frac{i_{ao}R_a}{v_{ao}}\right) = R_a \quad (3.63)$$

Stator and rotor windings thus have equal per-unit resistances which accurately reflect their mechanical identity. In fact, quite generally, with a physically meaningful base, the per-unit resistances of windings

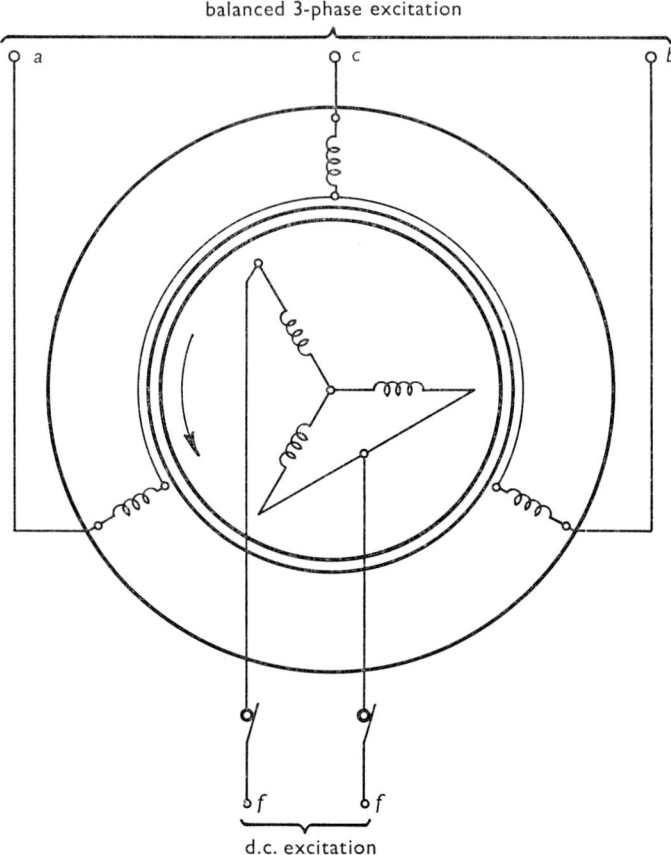

Fig. 12 Hypothetical synchronous induction motor in which stator and rotor windings are to be assumed identical

which perform the same function (i.e. are concerned with the same 'effect') within a given machine, can be seen to be roughly inversely proportional to their total volume of conductor material, due allowance being made for the size of overhangs, influences of shortchording and distribution etc. This meaningful comparison between windings within a machine disappears, however, with the irrational unit-voltage base.

3.3 Machines other than synchronous

The first two main Sections of this chapter have been related primarily to the synchronous machine. This is natural in view of the almost exclusive application of per-unit systems, in practice, to synchronous machines, and the simplification of the particularly complex rotor-circuit equations which arise for them. However, in the course of the discussion a number of principles have emerged, and it is instructive to consider how these might apply in the analysis of other machines. The aim is not to provide an exhaustive treatment, but to select a few points of interest.

3.3.1 Polyphase induction motor

In Section 3.2.1, the central importance of the fundamental sine wave of air-gap flux density for the basic performance of the synchronous machine, and for the definition of the (consequently physically meaningful) X_{ad} base, was illustrated. In the polyphase induction machine, the fundamental flux wave is of the same central importance, and it follows that the X_{ad} base is most natural and valuable for the definition of secondary rotor currents. In the same way that the coupling of this wave is ideal with the distributed polyphase armature of the synchronous machine, so the coupling in the induction machine is ideal with both the primary and secondary windings, these being of the same type (or effectively so in the case of a cage winding) as the synchronous-machine armature. Thus again, as in the synchronous machine, all other components of flux density, including the harmonic air-gap fluxes, can be treated as contributory to leakage reactance with some associated stray loss.† Such a classification is confirmed by Alger[30], for example.

Comparatively recently it has become important to examine the

† This involves some approximation, of course.

frequency response and stability characteristics of the 3-phase induction motor, and the 2-axis method of analysis has proved to be a powerful approach[35, 36]. Here, the 3-phase stator and rotor are both replaced by pseudostationary coils *ds* and *qs* for the stator, and *dr* and *qr* for the rotor, as in Fig. 13. The induction motor is, of course, nonsalient, and is completely symmetrical in both axes, and so, while

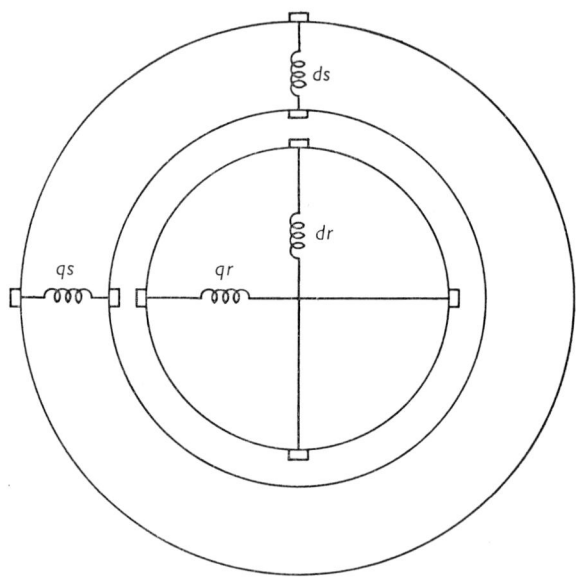

Fig. 13   3-Phase induction motor represented by pseudostationary 2-axis coils in both stator and rotor

we continue to refer to X_{ad} for convenience, it should be noted that, for both stator and rotor 3-phase coils,

$$\left.\begin{array}{l} X_{ds} = X_{qs} \\ X_{dr} = X_{qr} \\ X_{ads} = X_{aqs} \\ X_{adr} = X_{aqr} \end{array}\right\} \quad (3.64)$$

In many problems of interest, the zero-sequence variables may be assumed to be zero. A further consequence of the uniform air gap is that the m.m.f. base (Section 3.2.3) and X_{ad} base are identical. It remains true of course that it is the principle of the X_{ad} base which is meaningful.

The principles of the preferred fundamental 3-phase/2-axis per-unit system are entirely applicable to this problem. Thus, the transformation between stator per-unit 3-phase variables and *ds* and *qs* variables would have the form of eqns. 3.1–3.4. The same is true between rotor per-unit 3-phase variables and *dr* and *qr* variables, except that for the rotor transformation, assuming the reference frame fixed to the synchronously rotating field[35],

$$\theta_r = s\omega t + \delta \tag{3.65}$$

where *s* is fractional slip, in contrast to eqn. 3.6. As before, the voltage–current base product is $(3/2)v_{ao}i_{ao}$ in the 2-axis reference frame, and $v_{ao}i_{ao}$ in the 3-phase reference frame, peak values for base quantities again being preferred (see Section 4.1).

The base for the voltages v_{ds} and v_{qs} is v_{ao}, and that for the currents i_{ds} and i_{qs} is $(3/2)i_{ao}$. This is, of course, demanded by their relationship to the rated stator values. In choosing the base† for v_{dr} and v_{qr} and that for i_{dr} and i_{qr}, there is freedom to introduce a turns ratio, and it is at this stage that the X_{ad} principle is applied. Thus we may note that:

(a) Unit 2-axis rotor current (*dr* or *qr*) is the steady current which, flowing in the *dr* or *qr* coil, produces the same magnitude of fundamental air-gap flux wave as do rated, balanced stator currents flowing in the stator winding.

(b) Recalling the principle of equal effect embodied in the form of the preferred per-unit transformation equations, discussed in Section 3.1.1, it immediately follows that unit 3-phase rotor currents are automatically defined in the same manner; i.e. 'Unit-peak balanced 3-phase rotor currents flowing in the 3-phase rotor winding produce the same magnitude of fundamental air-gap flux wave as rated unit-peak balanced stator currents flowing in the stator winding.' It must of course be noted that the notion of 'the 3-phase rotor winding' may, in the case of a cage rotor, be simply an agreed basis for design calculations,

---

† In normal operation, the rotor is short-circuited and $v_{dr} = v_{qr} = 0$, but it should be appreciated that this does not affect the selection of their base quantity.

the 3-phase rotor currents having a known relationship to the actual bar currents.

For the present purpose, it is not necessary to show the exact form of the 2-axis equivalent circuits. It is sufficient to remark that, with this preferred choice of stator/rotor turns ratio, the per-unit series 'leakage' inductances of the ds and qs circuits and the dr and qr circuits are equal to the per-unit stator and rotor leakage inductances, respectively, as they would normally be calculated by the designer. Thus the equivalent circuits have their most physically meaningful form, as was the case for the synchronous machine (see Fig. 10).

The per-unit, per-phase equivalent circuit for balanced steady-state operation may be obtained from the 2-axis performance equations, by first simplifying the 2-axis equations so as to represent steady-state operation at a slip s, and then transforming the 2-axis solutions for $\boldsymbol{v}_{ds}$ and $\boldsymbol{v}_{qs}$ and $\boldsymbol{i}_{ds}$ and $\boldsymbol{i}_{qs}$ into the corresponding solutions for $\boldsymbol{v}_a$ and $\boldsymbol{i}_a$ (to take just one phase), by eqns. 3.3 and 3.4. It is not necessary to detail these steps, but the equivalent circuit which represents the outcome is of interest, and is shown in Fig. 14. As might intuitively be expected, the stator and rotor leakage reactances $\boldsymbol{X}_{ls}$ and $\boldsymbol{X}_{lr}$ in this well known form of phasor equivalent circuit are both the per-unit values corresponding to those obtained in the normal design calculation.

The benefit of an 'equal-effect' X_{ad} principle for the comparison of different windings may again be noted, in that if, hypothetically, the stator and rotor 3-phase windings were considered to be identical, one would obtain the physically meaningful result that

$$\left.\begin{array}{l} R_r = R_s \\ X_{lr} = X_{ls} \end{array}\right\} \quad (3.66)$$

With any other choice of base rotor currents, this would not be so. Because of the form of the preferred per-unit transformation equations, we have the following which is comparable to eqn. 3.8:

$$\left.\begin{array}{l} |\boldsymbol{I}_{as}| = \sqrt{(i_{ds}^2 + i_{qs}^2)} \\ |\boldsymbol{I}_{ar}| = \sqrt{(i_{dr}^2 + i_{qr}^2)} \\ |\boldsymbol{V}_a| = \sqrt{(v_{ds}^2 + v_{qs}^2)} \end{array}\right\} \quad (3.67)$$

The possibility of comparing an impedance value for the induction motor with that for the synchronous machine, for example, is of course an automatic consequence of employing a uniform per-unit system for both machines. Thus, assuming the stator winding to be capable of the same rated current and functioning as an armature in either type of machine, the per-unit stator resistance, and to a good approximation the leakage reactance, would be the same in both cases. This is not so if the principle of the per-unit system is varied, and in view of this, it should be noted that an alternative has been proposed[18] for the in-

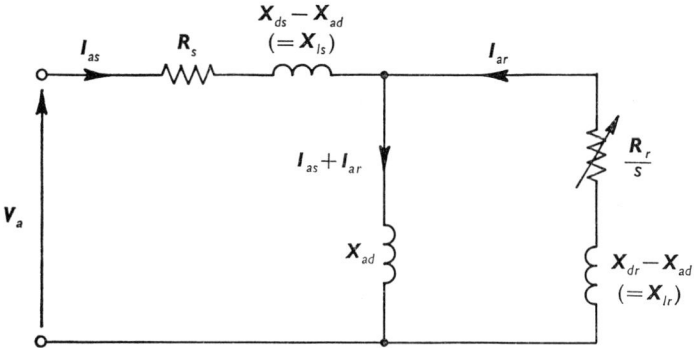

Fig. 14   Per-unit per-phase equivalent circuit of 3-phase induction motor in balanced steady-state operation using X_{ad} base

The core-loss resistance is neglected

duction motor, and is in fairly common use. This is that the base stator current should be taken as that corresponding to the in-phase component of the rated per-phase current. This choice has some practical merit; it avoids the base value depending on the rather uncertain design parameter of the full-load power factor, and it makes the base power equal to the total input (not apparent) electrical power. However, these advantages do not really compensate for the loss of fundamentality and uniformity of treatment, and the present authors would discourage its use. Stephen[5] supports this view.

3.3.2 Single-phase induction motor

The application of a per-unit system to the single-phase induction motor is of interest, not because of the importance of the application itself—since, to the authors' knowledge, it has received no direct attention—but for the opportunity it provides to demonstrate how the principles of earlier Sections are relevant to a significantly different machine. Attention is restricted to a simple single-phase motor with a uniform air gap and no starting devices (e.g. shaded pole). The former simplification is not important as far as the per-unit system is concerned, but it slightly reduces the complexity of the analysis. Such a machine is represented in Fig. 15. The polyphase rotor is adequately

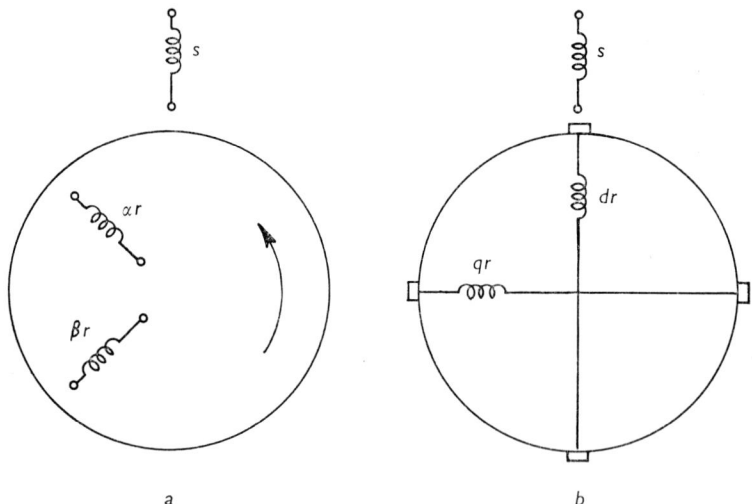

Fig. 15 Simple single-phase induction motor
a The single-phase stator and 2-phase rotor
b The rotor replaced by pseudostationary 2-axis coils

represented as 2-phase in Fig. 15*a*, and Fig. 15*b* shows the rotor replaced by equivalent pseudostationary 2-axis coils *dr* and *qr*.

Since the rotor is a 2-axis representation of a polyphase winding and the stator is a single-phase winding, the situation is essentially identical to that of a 2-phase synchronous machine with a single field winding, the roles of stator and rotor of course being inverted. The base voltage and current in the single-phase *s* coil are, in this case, automatically determined as the rated values. It is with the rotor 2-axis variables that the question of a choice of turns ratio arises, but the X_{ad} base is clearly to be preferred. The per-unit self inductances of coils dr and qr are equal to the per-unit synchronous inductance $\boldsymbol{L}_d$ of the α and β coils, while with this base, the mutual inductance between coils dr and s is equal to the per-unit armature-reaction inductance $\boldsymbol{L}_{ad}$ of the α and β coils.

In the per-unit system thus established, unit $\boldsymbol{i}_{dr}$ or $\boldsymbol{i}_{qr}$ produces the same fundamental air-gap flux-density wave as unit $\boldsymbol{i}_s$. The corresponding per-unit impedance matrix in the 2-axis reference frame can be written as

$$[\boldsymbol{Z}_{s,dr,qr}] = \begin{array}{c} s \\ dr \\ qr \end{array}\begin{array}{c} s \\ \left[\begin{matrix} \boldsymbol{R}_s+j\omega \boldsymbol{L}_s & j\omega \boldsymbol{L}_{ad} & \\ j\omega \boldsymbol{L}_{ad} & \boldsymbol{R}_r+j\omega \boldsymbol{L}_d & (1-s)\omega \boldsymbol{L}_d \\ (s-1)\omega \boldsymbol{L}_{ad} & (s-1)\omega \boldsymbol{L}_d & \boldsymbol{R}_r+j\omega \boldsymbol{L}_d \end{matrix}\right] \end{array} \quad (3.68)$$

This equation is restricted to steady-state operation so that inductive couplings can be represented in j notation and the rotor angular velocity is expressed in terms of slip s. It is implicit that all the voltage and current variables associated with $[\boldsymbol{Z}_{s,dr,qr}]$ are rated-frequency complex phasors. (It will be noted that the signs of the motional terms on the two axes are reversed, compared with those, for example, in eqn. 6.3. This is an unimportant consequence of the sign convention adopted. Comparing Fig. 15 with Fig. 4 for the 3-phase synchronous machine, it will be seen that the positive pseudomotion of the d and q coils relative to the single-phase coil is reversed in the two cases.)†

It must now be observed that, while there was a clear argument for the use of the X_{ad} base, reflecting the importance of the fundamental flux wave for the basic operation of the machine, an adequate understanding of the performance characteristics cannot be achieved without examining the significance of the base in respect of the two fundamental waves, one forward and one backward rotating, arising from the single-phase excitation of the stator. To facilitate this examination, a transformation from the dr and qr variables to new f and b variables

† The use of normalised angular velocity ω is considered in Section 4.3. It raises no problems for the present purpose, but reference may be made now to that Section if a need for clarification is felt.

is introduced.† It is defined as follows, its particular form being discussed later:

$$\begin{bmatrix} I_s \\ I_{dr} \\ I_{qr} \end{bmatrix} = \tfrac{1}{2} \begin{bmatrix} 2 & & \\ & 1 & 1 \\ & -j & j \end{bmatrix} \begin{bmatrix} I_s \\ I_f \\ I_b \end{bmatrix} \qquad (3.69)$$

the inverse being

$$\begin{bmatrix} I_s \\ I_f \\ I_b \end{bmatrix} = \begin{bmatrix} 1 & & \\ & 1 & j \\ & 1 & -j \end{bmatrix} \begin{bmatrix} I_s \\ I_{dr} \\ I_{qr} \end{bmatrix} \qquad (3.70)$$

The corresponding per-unit voltage transformation which maintains power invariance is determined by the counterpart of eqn. 2.29 for phasor variables[6]; i.e.

$$[V] = [C_t^*][V'] \qquad (3.71)$$

Thus

$$\begin{bmatrix} V_s \\ V_f \\ V_b \end{bmatrix} = \tfrac{1}{2} \begin{bmatrix} 2 & & \\ & 1 & j \\ & 1 & -j \end{bmatrix} \begin{bmatrix} V_s \\ V_{dr} \\ V_{qr} \end{bmatrix} \qquad (3.72)$$

the inverse being

$$\begin{bmatrix} V_s \\ V_{dr} \\ V_{qr} \end{bmatrix} = \begin{bmatrix} 1 & & \\ & 1 & 1 \\ & -j & j \end{bmatrix} \begin{bmatrix} V_s \\ V_f \\ V_b \end{bmatrix} \qquad (3.73)$$

The new impedance matrix is obtained by the counterpart of eqn. 2.32 for phasor variables; i.e.

$$[Z] = [C_t^*][Z'][C] \qquad (3.74)$$

which gives the result

$$[Z_{s,f,b}] = \begin{array}{c} s \\ f \\ b \end{array} \begin{bmatrix} R_s + j\omega L_s & j\omega \dfrac{L_{ad}}{2} & j\omega \dfrac{L_{ad}}{2} \\ js\omega \dfrac{L_{ad}}{2} & \dfrac{R}{2} + js\omega \dfrac{L_d}{2} & \\ j(2-s)\omega \dfrac{L_{ad}}{2} & & \dfrac{R_r}{2} + j(2-s)\omega \dfrac{L_d}{2} \end{bmatrix} \qquad (3.75)$$

Noting that for a cage rotor

$$V_f = V_b = 0 \qquad (3.76)$$

† It is possible to arrive at a per-phase equivalent circuit in contrarotating-field form without making this transformation[33], and indeed it is possible to obtain a per-phase equivalent circuit of a rather complex sort which does not appear to reflect the contrarotating fields at all[31], although of course it embodies the same information. However, the present approach is the most clear and straightforward.

it is very simple to restate the voltage/current relationships implied by eqn. 3.75, by multiplying throughout the second and third rows by $1/s$ and $1/(2-s)$, respectively, as

$$\begin{bmatrix} V_s \\ 0 \\ 0 \end{bmatrix} = \begin{bmatrix} R_s + j\omega L_s & j\omega \dfrac{L_{ad}}{2} & j\omega \dfrac{L_{ad}}{2} \\ j\omega \dfrac{L_{ad}}{2} & \dfrac{R_r}{2s} + j\omega \dfrac{L_d}{2} & \\ j\omega \dfrac{L_{ad}}{2} & & \dfrac{R_r}{2(2-s)} + j\omega \dfrac{L_d}{2} \end{bmatrix} \begin{bmatrix} I_s \\ I_f \\ I_b \end{bmatrix} \quad (3.77)$$

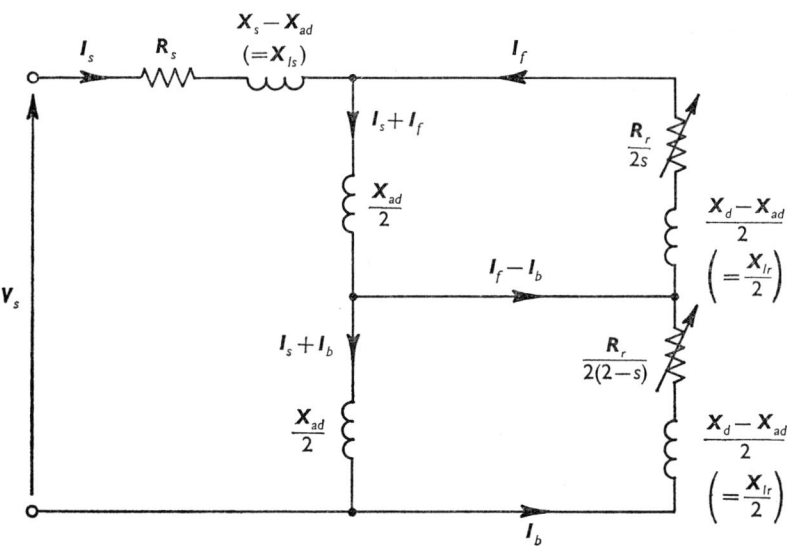

Fig. 16 Per-unit equivalent circuit of simple single-phase induction motor, assuming preferred per-unit transformation equations (eqns. 3.69–3.73)

By inspection of eqn. 3.77, it is seen that a single equivalent circuit can be drawn in the form of Fig. 16. The form of this equivalent circuit is common in the literature[32,33], though it has not been discussed from the viewpoint of a per-unit system.

This equivalent circuit has the desirable property of making the stator series 'leakage' reactance equal to $X_s - X_{ad}$, which, as was noted in studying the synchronous machine and the polyphase induction machine, is equal to the per-unit stator leakage reactance normally calculated by the designer. The corresponding per-unit leakage reactance $X_d - X_{ad}$ of the rotor is seen to be associated half with the forward wave and half with the backward wave. The per-unit resistance R_r of the rotor is divided in the same way. It must be noted, however, that these results are dependent on the particular choice of transformation equations.

If, for example, the transformations are made orthogonal,† an equivalent circuit is obtained having a much less useful form in per-unit terms. The per-unit series reactance of the stator becomes $X_s - \sqrt{2}(X_{ad})$, which is totally unrelated to the per-unit stator leakage reactance, and can easily be negative. All the other reactances in the equivalent circuit become less meaningful, while the per-unit resistance associated with each wave becomes R_r, instead of $R_r/2$. This approach is demonstrated by the treatment in Reference 34, which is, however, in ordinary variables, and is not concerned with a per-unit system.

There is thus a similarity to the case of the 3-phase/2-axis per-unit system, in the respect that the orthogonal transformation of per-unit variables again does not represent a preferred choice on grounds of physical meaningfulness. In another respect, however, there is an interesting difference: in the present case, it is not possible to introduce a different voltage–current base product for the f, b variables compared with the dr, qr variables; this is in contrast to the preferred treatment of the d, q, z and a, b, c variables. The reason—the principle of which was mentioned in the discussion of Section 3.1.1—is that the s circuit is common to both the (s, dr, qr) and (s, f, b) reference frames, and both reference frames are useful in analysis. It follows that the s, dr, qr, f and b variables must all share a common voltage–current base product to produce a useful and physically meaningful overall per-unit system. This illustrates nicely the point made in Section 3.1.2 that the establishment of a per-unit system for a par-

---

† This is achieved by making the coefficient of eqn. 3.69 equal to $1/\sqrt{2}$ instead of $\tfrac{1}{2}$, and modifying the element 2 within the matrix to $\sqrt{2}$, so that I_s remains unchanged.

ticular field of analysis has to be considered on its merits, using broad guiding principles.†

There are, therefore, reasons for preferring eqns. 3.69–3.73 as transformation equations for per-unit variables. It is profitable to consider the manner in which the broad principles discussed in Section 3.1.2 can be seen to apply to this choice, (*b*) of that Section—the principle of equal effect—being relevant.

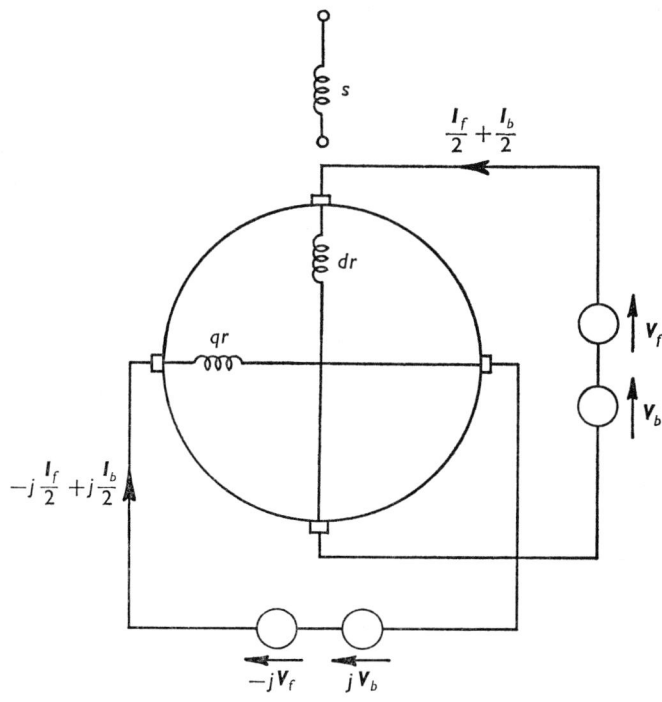

Fig. 17 Components of current and voltage flowing in the pseudostationary *dr* and *qr* coils, in accordance with the preferred per-unit transformation equations for the single-phase induction motor.

The physical interpretation of the above equations —as for the symmetrical components of Section 3.1.4—is that the *f*, *b* variables are

† It should not be inferred, incidentally, that a varied voltage–current base product would necessarily be an advantage in the present case, if it were available. The point is simply that it cannot be considered.

components of the *dr*, *qr* variables, applied to the *dr*, *qr* coils. The preferred choice of the components leads to the situation shown in Fig. 17, which reveals clearly that the *f*, *b* variables represent two 2-phase excitation systems with oppositely directed phase sequences which therefore excite oppositely rotating fields. The statement of their equivalence of effect to the *dr*, *qr* variables, and hence to the *s* variables, has to be made with care. Considering, for illustration, that $\boldsymbol{I}_{dr} = 1$, and $\boldsymbol{I}_{qr} = 0$, it follows from eqn. 3.70 that

$$\boldsymbol{I}_f = \boldsymbol{I}_b = \boldsymbol{I}_{dr} \qquad (3.78)$$

and from this we can deduce that unit $\boldsymbol{I}_f$ or $\boldsymbol{I}_b$ represents a set of balanced 2-phase currents which, flowing alone in the *dr*, *qr* coils, produce a rotating fundamental flux wave of the same peak value as the peak of the alternating fundamental wave produced by unit $\boldsymbol{I}_{dr}$ flowing in the *dr* coil. Hence, by the X_{ad} principle relating $\boldsymbol{I}_{dr}$ to $\boldsymbol{I}_s$, the rotating fundamental flux wave has the same peak value as that of the alternating fundamental flux wave produced by unit $\boldsymbol{I}_s$ flowing in the *s* coil.

It will be clear that, with, for example, the changed factor of $1/\sqrt{2}$ which is introduced by the orthogonal transformations, no such equal-effect principle for unit currents would be recognisable.

3.3.3 Commutator machines

It is not the intention to pursue here the matter of establishing preferred per-unit systems for the various types of commutator machine, but simply to note briefly the nature of a problem that could arise in regard to the specification of secondary-current bases. It may be noted that this area of application for per-unit systems is relatively unexplored at the present time.

With regard to a.c. commutator machines operating basically on the induction-motor principle—for example, the Schrage motor—it is clear that the fundamental air-gap flux wave is again basic to the stator/rotor interaction. Accordingly, the X_{ad} base represents the natural and physically meaningful principle for the evaluation of secondary-current bases; the problem of defining bases for the various circuits is essentially the same as for the induction and synchronous machines discussed earlier, and merits no further discussion here.

With regard to d.c. commutator machines, the situation is more complicated since the odd harmonics of the air-gap flux density specifically contribute to the motional e.m.f. produced at the brushes, and hence take part in the basic mode of operation of the machine. Clearly, therefore, the X_{ad} principle is irrelevant for this case.

It would be possible to avoid considering the basic problem by choosing to normalise both the armature and field circuits to the rated terminal values of voltage and current. This of course implies unit turns ratio between the two circuits. Such a definition would have some practical usefulness; e.g. the per-unit field current for the shunt-connected machine, and conversely the field voltage for the series-connected one, would equal the proportion of rated power dissipated as ohmic loss in the field circuit. On the other hand, no meaningful principle of equal effect has been invoked.

An alternative equal-effect definition would be that base field current is defined to produce the same motional e.m.f. between the quadrature-axis brushes of the armature, at a given speed, as rated armature current flowing between the hypothetical direct-axis brushes. However, having drawn attention to these two possibilities, it may be remarked that the subject is a complex one requiring an extensive discussion, which will not be given here.

4 MECHANICAL AND OTHER CONSIDERATIONS

4.1 Mean and instantaneous power; r.m.s. and peak bases

As mentioned in Section 1.4, the base quantities of alternating voltage and current are sometimes taken as the r.m.s. values[6,8,16,18] and sometimes as the peak values[2,7,10,11]. The latter definition is strongly recommended here for the fundamental 3-phase/2-axis per-unit system, for reasons which do not appear to have been fully stated previously, and which are accordingly given below.

If the r.m.s. values are assumed, then the per-unit peak values of balanced, rated, 3-phase voltages and currents are $\sqrt{2}$. Consequently, the peak flux linkage with any phase coil in which the rated voltage is being induced is $\sqrt{2}$ (see Section 4.3), and under rated load

$$\sqrt{(v_d^2 + v_q^2)} = \sqrt{(i_d^2 + i_q^2)} = \sqrt{2}$$

Similarly, a current of $\sqrt{2}$ is required in a rotor coil to produce 'equal effect'—to produce the same fundamental flux-density wave—to that of the rated 3-phase currents in the armature winding. It is surely preferable to have unity rather than $\sqrt{2}$ associated with these rated conditions, thus conforming to the principles of Section 3.1.2. This result is obtained by taking peak values as the base quantities.

The opposing argument is that r.m.s. values are of more common occurrence in a.c. theory. The answer to this is that a detailed 2-axis analysis, on the lines which have been discussed, generally implies an interest in transient problems, in which the need for care in discriminating between peak and r.m.s. is already well known. The peak value is the more significant in such work and unity may sensibly be assigned to it.

It might also be argued that the mean apparent power in an a.c. circuit with unit impressed voltage and current is 0·5 by the peak system, but, more conveniently, unity by the r.m.s. system. (Such an argument is not strictly complete, as there is no essential requirement for the power base of a per-unit system to equal the voltage–current base product. In the system under discussion it is so for 2-axis variables, but not for per-phase variables.) But, in transient problems, mean power is not relevant; the expressions for instantaneous power are those of interest. The form of these expressions is, overall, equally convenient in either system, as will be shown in eqns. 4.4 and 4.6.

The calculation of mean power implies a steady-state problem, and here one might define a completely new per-unit system, or consider the use of 'associated per-unit parameters'. This concept will be considered in the next Section. It represents a simple rationalisation of procedures implicit in normal practice. The fundamental base quantities are therefore taken here as the peak rated voltage and current per phase of the armature, and the phases may usefully be defined as equivalent star-connected phases, irrespective of the manner of connection of the real phase windings. This simple standardisation agrees with the recommendation of Reference 18. It is of limited importance —for example, the form of the per-unit equations is totally unaffected —but it ensures a constant relationship between the actual 3-phase and base quantities; thus, rated line voltage = $\sqrt{3}$ × base voltage, and rated line current = base current. In addition, the fundamental base of power is defined as the total rated apparent power, which is also the 2-axis voltage–current base product, so that

$$p_0 = (3/2)\, v_{ao} i_{ao} \tag{4.1}$$

Consider now the expression for instantaneous ordinary 3-phase power

$$p = v_a i_a + v_b i_b + v_c i_c \tag{4.2}$$

Rewriting this equation, prior to normalising to bases p_o, v_{ao} and i_{ao}, yields

$$\frac{p}{\tfrac{3}{2} v_{ao} i_{ao}} = \frac{2}{3}\left(\frac{v_a i_a}{v_{ao} i_{ao}} + \frac{v_b i_b}{v_{ao} i_{ao}} + \frac{v_c i_c}{v_{ao} i_{ao}}\right) \tag{4.3}$$

and hence the equation for instantaneous per-unit 3-phase power is

$$p = 2/3(v_a i_a + v_b i_b + v_c i_c) \tag{4.4}$$

The instantaneous ordinary 2-axis power is

$$p = v_d i_d + v_q i_q + 2 v_z i_z \tag{4.5}$$

It will be recalled from Section 3.1.1 that the base value of a 2-axis current is $(3/2)i_{ao}$, so that eqn. 4.5 normalises very simply, to become

$$p = v_d i_d + v_q i_q + 2 v_z i_z \qquad (4.6)$$

It may be remarked that the anomalous factor of 2 in the zero-sequence component of power arises from the non-power-invariant—though practically acceptable—transformation of zero-sequence parameters, mentioned in Section 3.1.1. The difference in the overall coefficients—2/3 in eqn. 4.4 and unity in eqn. 4.6—is a consequence of the difference in voltage–current base products between the two reference frames. It is a price paid for the advantages of the varied base product.

In literature which uses r.m.s. values for base quantities, the right-hand side of eqns. 4.4 and 4.6 are both modified by a factor of 1/2.

Having defined the independent bases of the fundamental 3-phase/2-axis per-unit system, the particular parameters on which the bases were established—peak phase voltage, peak phase current and total apparent power—may be usefully referred to as the 'fundamental base parameters'. The bases so established are appropriate for the detailed study of many machine problems, but other definitions can usefully be made. These are discussed in the next Section.

4.2 Concept of associated per-unit parameters

The discussion in the previous Section has continued the study of the fundamental 3-phase/2-axis system, and it should be said that this is a system which, with some variations, has found application in many problems of machine analysis. Examples are the transient behaviour of synchronous generators, the asynchronous run-up characteristics of synchronous motors, and, more recently, the stability characteristics of polyphase induction machines. These problems share the property—as was mentioned in Section 1.2—that they represent detailed analyses of the internal characteristics of machines. For problems more concerned with the power-supply system as a whole, and less with the individual machines, other base parameters are used in per-unit systems. In Section 2.1, for example, it was mentioned that, for the steady-state analysis of the single-phase transformer, it would be normal to define base voltages and currents as r.m.s. values. Equally, in Section 3.1.4, which discusses phasor symmetrical components, the base voltages and currents would again normally be r.m.s. values; moreover, in 3-phase analysis, it is quite common to define base values of line voltages and phase voltages[11] for example.

For each different area of analysis, a simple view would be that a new and totally separate per-unit system has been defined, and this is certainly a workable assumption. However, the purpose of this Section is to illustrate briefly that it is possible alternatively to regard some per-unit systems as being interrelated, and to consider them in a comprehensive manner. It is thought that this may be a matter of some general interest.† Additionally, it can be helpful in the discussion of certain problems where areas of application of different per-unit systems overlap; e.g. in the detailed analysis of a machine subject to unbalanced supply voltages from a larger system.

The essential point is that, having defined a fundamental per-unit system, established on a particular choice of fundamental base parameters, it is possible to define with discretion what may be termed 'associated per-unit parameters'. Each associated parameter is normalised to a different base, the principle of definition being that each one becomes unity under the same conditions that make the corresponding fundamental per-unit base parameter unity. Thus, for example, the associated per-unit r.m.s. line voltage can be defined, for balanced conditions, to be unity when the system voltage has rated value. Its base quantity is readily seen to be $v_{ao}\sqrt{(3/2)}$. When, for example, the fundamental per-unit peak phase voltage is unity, the fundamental per-unit peak line voltage is $\sqrt{3}$, while the fundamental r.m.s. line voltage is $\sqrt{(3/2)}$. However, correspondingly, the 'associated' peak line voltage is unity (being normalised to $\sqrt{3}v_{ao}$) and the 'associated' r.m.s. line voltage is also unity [being normalised to $v_{ao}\sqrt{(3/2)}$].

The advantage of this nomenclature is that it interconnects many of the per-unit parameters used variously with machines and power systems, and it logically relates them to a set of fundamental per-unit

† It is included for this reason, and should not be taken to be an established approach to the subject.

parameters. The system would be tedious if it were always necessary to describe fully every parameter as 'associated' or 'fundamental', but, in practice, the full description is often unnecessary, since it is clear from the nature of the problem under discussion.

As an example of the use of associated per-unit parameters, the form of equation for the mean total power in a balanced polyphase system with any number of phases is

$$P = \text{Re } \boldsymbol{VI}^* \qquad (4.7)$$

Here, $\boldsymbol{V}$ and $\boldsymbol{I}$ are the associated per-unit r.m.s. line or phase voltages and currents.

4.3 Flux linkage, instantaneous torque and per-unit time

It has been argued[8] that it is an unnecessary complication to normalise time in the fundamental 3-phase/2-axis per-unit system. It is, however, common in American literature [2,3,10,11,16,21,27–29], and the present authors prefer it for the reasons given below.

The per-unit system is simplified, and agrees better with the first of the principles of Section 3.1.2, if the per-unit electrical angular velocity ω_0 under rated (full load) conditions is made equal to unity. This is of little importance in itself for a fixed-frequency machine, but is very convenient when the frequency is a variable, and it has become significant with the increasing use of variable-frequency supplies to synchronous and induction machines. A further advantage is that the form of the expressions for torque become more appropriate, and this is shown later in this Section (eqn. 4.20).

Since ωt is a fundamental dimensionless group, the selection of ω_0 as a base of ω is equivalent to selecting $1/\omega_0$ as a base of time (i.e. $t_0 = 1/\omega_0$). Consider, for example, the normalisation of the equation for the instantaneous voltage of phase a:

$$v_a = \hat{v} \sin \omega t \qquad (4.8)$$

Rewriting this equation prior to normalising the variables gives

$$\frac{v_a}{v_{ao}} = \frac{\hat{v}}{v_{ao}} \sin\left(\omega_0 t_0 \frac{\omega}{\omega_0} \frac{t}{t_0}\right)$$

and it clearly follows that this only assumes the simple per-unit form

$$v_a = \hat{v}\sin\omega t \qquad (4.9)$$

when

$$t_o = \frac{1}{\omega_o} \qquad (4.10)$$

It may here be noted that, since reactance is the product of inductance and angular velocity, and the latter is unity at rated frequency, the per-unit inductance and per-unit rated-frequency reactance are equal. It is therefore common to find in the literature no distinction drawn between these quantities *where time is normalised*, as in Reference 29, for example.

The normalisation to form per-unit impedances was discussed in Section 3.1.1, but the manner of normalisation of the differential operator was not discussed at that stage. Having normalised time, it is now possible to see the consequent definition of **p**, the per-unit differential operator. Thus, in ordinary form,

$$p = \frac{\partial}{\partial t} \qquad (4.11)$$

but $t = t/\omega_o$ from eqn. 4.10; so it follows that

$$\mathbf{p} = \frac{\partial}{\partial t} = \frac{1}{\omega_o}\frac{\partial}{\partial t} = \frac{p}{\omega_o} \qquad (4.12)$$

The exact form of the operational per-unit impedances quoted in Section 3.1.1 may now be noted; e.g. eqn. 3.26 is

$$\mathbf{L}_{df}\mathbf{p} = N_f\left(\frac{3}{2}\frac{i_{ao}}{v_{ao}}\right)(\omega_o L_{afd})\left(\frac{p}{\omega_o}\right) = N_f\left(\frac{3}{2}\frac{i_{ao}}{v_{ao}}\right)(X_{afd})\left(\frac{p}{\omega_o}\right) = \mathbf{X}_{df}\mathbf{p} \qquad (4.13)$$

confirming that $\mathbf{X}_{df}$ and $\mathbf{L}_{df}$ are numerically equal.

Another dependent base which is determined as a consequence of normalising the angular frequency is that of flux-linkage. For example, consider Park's equations for the direct-axis, quadrature-axis and zero-sequence variables; these may be written in ordinary form as

$$\left.\begin{array}{l} v_d = p\psi_d - \psi_q p\theta + i_d R_d \\ v_q = p\psi_q + \psi_d p\theta + i_q R_q \\ v_z = p\psi_z + i_z R_z \end{array}\right\} \qquad (4.14)$$

It will be appreciated that eqns. 4.14 are no more than a restatement, in alternative form, of information contained in the equivalent circuits in Fig. 5, making use of the general expression for flux linkage given by eqn. 3.33. The equation for d-axis voltage may be written in preparation for normalisation as

$$\left(\frac{v_d}{v_{ao}}\right) = \left(\frac{p}{\omega_o}\right)\left(\frac{\omega_o}{v_{ao}}\psi_d\right) - \left(\frac{\omega_o}{v_{ao}}\psi_q\right)\left(\frac{p\theta}{\omega_o}\right) + \left\{\frac{i_d}{(3/2)i_{ao}}\right\}\left(\frac{3i_{ao}}{2v_{ao}}R_d\right)$$

so that, noting the nature of $\boldsymbol{R}_d$ from eqn. 3.28, the per-unit equation is

$$v_d = p\psi_d - \psi_q p\theta + i_d R_a$$

and the dependent form of the flux-linkage base is

$$\psi_o = \frac{v_{ao}}{\omega_o} \qquad (4.15)$$

It is clear from eqn. 4.15 that unit flux linkage ψ_o is the peak flux linkage which, varying at rated frequency, induces rated voltage; in other words, it is the peak flux linkage occurring in a phase coil, which has rated voltage induced in it on open-circuit. It therefore agrees with the first principle of Section 3.1.2, but this would not be so if time were not normalised. The normalised form of eqns. 4.14 may now be noted:

$$\left.\begin{aligned} v_d &= p\psi_d - \psi_q p\theta + i_d R_a \\ v_q &= p\psi_q + \psi_d p\theta + i_q R_a \\ v_z &= p\psi_z + i_z R_a \end{aligned}\right\} \qquad (4.16)$$

When the complete set of machine equations (i.e. circuit and torque equations) are written in ordinary parameters, it is inevitable that the number of pole pairs n appears. It is desirable in forming the per-unit equations to remove this parameter, both because it is not fundamental to the behaviour of the machine and because as a consequence the form of the per-unit angular velocity of the rotor is simplified, becoming the same whether expressed in mechanical or electrical form. This is done by defining the base mechanical angular velocity ν_o as that corresponding to base electrical velocity ω_o. Thus

$$\nu_o = \frac{\omega_o}{n}$$

whence
$$\boldsymbol{\nu} = \frac{\nu}{\nu_o} = \frac{n}{\omega_o}p\theta_m = \frac{p\theta}{\omega_o} = \mathbf{p}\theta \qquad (4.17)$$

This may be thought of as normalising the mechanical angle θ_m to a base of $1/n$ rad, normalising here simply implying division by a characteristic value.

The base torque then follows by dependence, as that corresponding to base power at base mechanical speed; i.e.

$$T_o = \frac{\frac{3}{2} v_{ao} i_{ao}}{v_o} \qquad (4.18)$$

It may here be noted that Adkins[8] proposes an alternative system, in which torque and mechanical speed of a multipole machine are normalised to the bases of an equivalent 2-pole machine of the same power. This complication seems to offer little advantage.

The self consistency of the above definitions may now be exemplified by deriving the per-unit equation for torque in terms of flux linkage and current. The ordinary 2-axis electrical power associated with the motional terms of eqns. 4.14 (equivalent to mechanical output shaft power) may be written as

$$p = Tv = \psi_d i_q \mathrm{p}\theta - \psi_q i_d \mathrm{p}\theta \qquad (4.19)$$

Rearranging this equation in preparation for normalising yields

$$\left(\frac{v_o T}{\frac{3}{2} v_{ao} i_{ao}}\right)\left(\frac{v}{v_o}\right) = \left(\frac{\psi_d \omega_o}{v_{ao}}\right)\left(\frac{i_q}{\frac{3}{2} i_{ao}}\right)\left(\frac{\mathrm{p}\theta}{\omega_o}\right)$$

$$- \left(\frac{\psi_q \omega_o}{v_{ao}}\right)\left(\frac{i_d}{\frac{3}{2} i_{ao}}\right)\left(\frac{\mathrm{p}\theta}{\omega_o}\right)$$

Noting the results for eqns. 4.17 and 4.18 and introducing per-unit symbols gives the per-unit equation for instantaneous torque:

$$\boldsymbol{T} = \boldsymbol{\psi}_d \boldsymbol{i}_q - \boldsymbol{\psi}_q \boldsymbol{i}_d \qquad (4.20)$$

This physically significant simple form of the torque equation is only obtained when time is normalised. If this is not done, the inappropriate factor ω_o appears in the equation (see, for example, Reference 8).

4.4 Inertial torque

The ordinary equation for torque, in terms of moment of inertia J and mechanical angular acceleration, is

$$T = J\mathrm{p}v \qquad (4.21)$$

Rearranging this equation in preparation for normalising gives

$$\left(\frac{v_o T}{\frac{3}{2}v_{ao}i_{ao}}\right) = \left(\frac{\omega_o v_o^2 J}{\frac{3}{2}v_{ao}i_{ao}}\right)\left(\frac{\mathbf{p}}{\omega_o}\right)\left(\frac{v}{v_o}\right) \quad (4.22)$$

This equation reduces to its most widely used form (see, for example, References 8 and 11), when expressed in terms of the so-called 'inertia constant' H, where H is the ratio of stored mechanical energy at unit mechanical speed to rated apparent power.

Therefore

$$H = \frac{\frac{1}{2}Jv_o^2}{\frac{3}{2}v_{ao}i_{ao}} \quad (4.23)$$

and eqn. 4.22 becomes $\quad T = 2\omega_o H\mathbf{p}v = 2\omega_o H\mathbf{p}^2\theta \quad (4.24)$

The complicated coefficient $2\omega_o H$ in eqn. 4.24 is inappropriate. It neither simplifies the form of the equation, nor does it parallel the nature of the ordinary analysis. It arises through the use of H, which has the dimensions of time in seconds. The present authors favour Concordia's use[10] of 'per-unit moment of inertia', which is defined as

where
$$\left. \begin{array}{l} \mathbf{J} = \dfrac{J}{J_o} = 2\omega_o H \\[6pt] J_o = \dfrac{3}{2}\dfrac{v_{ao}i_{ao}}{\omega_o v_o^2} \end{array} \right\} \quad (4.25)$$

so that eqn. 4.24 is reduced to its simplest form:

$$\mathbf{T} = \mathbf{J}\mathbf{p}^2\theta \quad (4.26)$$

Stephen[5] refers to both H and $\mathbf{J}$, but prefers to call $\mathbf{J}$ the 'inertia constant' and H the 'stored-energy constant'. This may be a sensible nomenclature, but it is unfortunately likely to cause confusion. The longer term 'per-unit moment of inertia' is consequently preferred here for $\mathbf{J}$.

Physically, $\mathbf{J}$ may be thought of as

$$\mathbf{J} = 2 \times \frac{\text{stored mechanical energy at unit mechanical speed}}{\text{rated apparent-energy input in unit time}} \quad (4.27)$$

Of course, unit moment of inertia is that which has unit acceleration under unit torque.

5 RECOMMENDATIONS AND CONCLUSIONS

The main ideas and results which have emerged from the discussion in the previous chapters are briefly reviewed here to provide a more concise and clear overall picture of per-unit systems, and it is hoped, at the same time, to illuminate some ideas rather more fully as a result of their juxtaposition with others.

Much of the earlier discussion has been concerned with the preferred form of a fundamental 3-phase/2-axis per-unit system, and considerable emphasis has been laid throughout, not only upon sound mathematical formulation, but also particularly on the physical aspects of the subject as a whole. The system developed provides the basis for the analysis of a very wide range of different devices and systems, and applies to virtually the entire field of machine problems to which per-unit systems have been previously applied. In the past, the most important area of application has been to the circuit problems of the 3-phase synchronous machine, but, more recently, attention has been given to studies of 3-phase induction machines. In the future, there will inevitably be a substantial increase in the application of per-unit systems, not only to synchronous machines and systems, but also to induction machines, particularly in view of the increasing complexity of systems generally and the associated increase in importance of transient behaviour. To this work the use of per-unit systems will bring the advantages discussed fully in Chapter 1.

The features which together define the form of the 'preferred' system recommended by the authors are grouped below, each being followed by a reference to the Section(s) in which it is primarily discussed.

    (*a*) The transformation equations interrelating per-unit 3-phase and 2-axis variables have the form of eqns. 3.1 and 3.2, with the inverse form of 3.3 and 3.4. The advantages of this choice spring basically from the unit–unit relationship which is thereby established between variables of the two reference frames (Section 3.1.1).

(b) The base for secondary currents is that which is here called the 'full-pitch X_{ad}' base, and this is the one having greatest physical significance. With this choice, the base secondary currents in distributed polyphase windings, or in windings which are full-pitched (or replaceable by equivalent full-pitched windings), are defined so as to produce the same fundamental air-gap flux wave as balanced, rated currents in the primary windings. Windings which are effectively short-pitched (as is the case in most damper circuits) are referred by a 1:1 ratio to the equivalent full-pitched winding (Sections 3.2.1 and 3.2.2).

(c) The base primary 3-phase voltage v_{ao} and current i_{ao} are defined as the peak rated values for an equivalent star-connected machine (Section 4.1). The corresponding bases in the 2-axis reference frame are v_{ao} and $\frac{3}{2}i_{ao}$ (Section 3.1.1).

(d) The base of power equals the total rated apparent power equals $(3/2)v_{ao}i_{ao}$ (Section 4.1).

(e) Time is normalised to a base $1/\omega_o$, where ω_o is the rated angular frequency of the machine (Section 4.3).

(f) Mechanical angular velocity is normalised, so that rotor speed, for example, is the same in mechanical or electrical terms (Section 4.3).

(g) The use of per-unit moment of inertia is recommended; the inertia constant being a nonpreferred alternative (Section 4.4).

In the development and study of the above system, a number of points of broader interest have emerged and some of the more important groups of these are noted below.

(i) The value of a per-unit system and the understanding of it (e.g. through the use of equivalent circuits) are increased if its application to a particular device is understood in physical, as well as purely mathematical, terms. This objective is promoted by clearly separating the following three steps in establishing the system:

transformation of the variables

normalisation of the complete set of variables in a single operation

application of wisely chosen turns ratios to the relevant circuits.

(ii) The numerical values ascribed to the interrelated concepts of turns ratio and leakage reactances associated with individual windings are fundamentally indeterminate and of no physical significance in general. An ideal ratio can be defined with great care, which, in highly restricted circumstances, does have physical significance. Fortunately, because of the nature of most practical devices, an ideal ratio may be, and generally has been, assumed to exist. The full-pitch X_{ad} base is defined through the recognition of such a ratio for devices in which the fundamental sine wave of air-gap flux is completely basic to their behaviour. It is also an example of the application of the principle of equal effect, which is extremely valuable in constructing any per-unit system model of a physically significant kind.

(iii) The preferred transformation of per-unit variables from the 3-phase to the 2-axis reference frames is derived, so that, as far as possible, unity among per-unit variables is associated with the condition of rated load in the two reference frames simultaneously. This is achieved by varying the voltage–current base product in the two reference frames. This technique cannot be applied in all analyses, and it emphasises the need for flexibility in constructing a per-unit system in the light of particular circumstances. It is not implied that the base of power is varied, the form of the per-unit power equations being appropriately adjusted. The effect is that 3-phase windings on one member are treated as a single unit as far as the per-unit system is concerned, thus effectively achieving the one-to-one correspondence between per-unit per-phase impedances and 2-axis impedances, which occurs automatically with a 2-phase machine.

The usefulness and range of applicability of both the recommended procedure for the construction of a per-unit system and some of the broader ideas just considered have been demonstrated by the treatment of the polyphase induction machine, which is quite straightforward, more effectively by the treatment of the simple single-phase induction machine and, finally of commutator machines. It is believed that the treatments of the last two machine classes are the first to be presented, and that that for the single-phase machine, in terms of the X_{ad} base, is complete. Treatment is also complete for most a.c. commutator machines, but, for the d.c. machine, comment is restricted to noting the differences between that and the a.c. machine and to proposing a possible equal-effect principle.

In conclusion, it may be noted that the reason which caused the authors to consider together the subject of per-unit systems, particularly as they apply to machines, was the awareness of the confusion caused by the widely varying use and misuse of such systems in the literature. The present Monograph has attempted to clarify and correlate earlier literature, but, more important, it has argued the case for a particular preferred system. The authors hope that this system will be accepted as the agreed standard from now on. Finally, it cannot be overemphasised that whatever per-unit system is used, it is essential that a distinct notation is employed between ordinary and per-unit equations, and that a clear and complete statement is included which defines, unambiguously, the system and its relationship to others.

6 APPENDIXES

6.1 Transformation of impedance matrix from 3-phase to 2-axis parameters

In the following transformation, one rotor circuit (f) is included to demonstrate typical behaviour. The transformation is made in ordinary variables, using the transformation equations (eqns. 3.14 for voltage, and eqns. 3.20 for current). It should be noted that this is not an orthogonal transformation, and that the transformation of zero-sequence variables is not power-invariant.

In the (a, b, c, f) reference frame, the voltage/current relationships may be written as

$$\begin{bmatrix} v_a \\ v_b \\ v_c \\ v_f \end{bmatrix} = \begin{bmatrix} R_a + pL_{aa} & pL_{ab} & pL_{ca} & pL_{af} \\ pL_{ab} & R_a + pL_{bb} & pL_{bc} & pL_{bf} \\ pL_{ca} & pL_{bc} & R_a + pL_{cc} & pL_{cf} \\ pL_{af} & pL_{bf} & pL_{cf} & R_f + L_{ff}p \end{bmatrix} \begin{bmatrix} i_a \\ i_b \\ i_c \\ i_f \end{bmatrix} \quad (6.1)$$

The sign convention treats both stator and field as loads; i.e. stator and field currents are inward-directed, flowing in the direction of applied terminal voltage.

With the usual idealising assumptions about the variation of inductances with θ, we may write

$$\left. \begin{aligned} L_{aa} &= L_{ao} + L_{a2} \cos 2\theta \\ L_{bb} &= L_{ao} + L_{a2} \cos(2\theta - 240°) \\ L_{cc} &= L_{ao} + L_{a2} \cos(2\theta - 120°) \\ L_{ab} &= -L_{abo} + L_{a2} \cos(2\theta - 120°) \\ L_{bc} &= -L_{abo} + L_{a2} \cos 2\theta \\ L_{ca} &= -L_{abo} + L_{a2} \cos(2\theta - 240°) \\ L_{af} &= L_{afd} \cos \theta \\ L_{bf} &= L_{afd} \cos(\theta - 120°) \\ L_{cf} &= L_{afd} \cos(\theta - 240°) \end{aligned} \right\} \quad (6.2)$$

where L_{ao}, L_{a2}, L_{abo} are constant coefficients (the subscript o does not here signify a base quantity), and L_{afd} is as defined elsewhere.

After transformation, the voltage/current relationships in the (d, q, z, f) reference frame are found to be

$$\begin{bmatrix} v_d \\ v_q \\ v_z \\ v_f \end{bmatrix} = \begin{bmatrix} R_d + L_{dd}\mathrm{p} & -L_{qq}(\mathrm{p}\theta) & & L_{df}\mathrm{p} \\ L_{dd}(\mathrm{p}\theta) & R_q + L_{qq}\mathrm{p} & & L_{df}(\mathrm{p}\theta) \\ & & R_z + L_{zz}\mathrm{p} & \\ L_{df}\mathrm{p} & & & R_f + L_{ff}\mathrm{p} \end{bmatrix} \begin{bmatrix} i_d \\ i_q \\ i_z \\ i_f \end{bmatrix} \quad (6.3)$$

The electrical angle θ is measured between the axis of phase a and the direct axis (i.e. the axis of the field) and $(\mathrm{p}\theta)$ is positive when the field rotates in the direction $a \to b \to c$.

The following relationships are shown to exist by the transformation:

$$\left.\begin{aligned} R_d &= R_q = \tfrac{2}{3} R_a \\ L_{dd} &= \tfrac{2}{3}(L_{ao} + L_{abo}) + L_{a2} \\ L_{qq} &= \tfrac{2}{3}(L_{ao} + L_{abo}) - L_{a2} \\ L_{df} &= L_{afd} \\ R_z &= \tfrac{2}{3} R_a \\ L_{zz} &= \tfrac{2}{3}(L_{ao} - 2L_{abo}) \end{aligned}\right\} \quad (6.4)$$

When normalised to the bases of the fundamental 3-phase/2-axis per-unit system, recalling that the bases for the a, b and c parameters are two-thirds of those for the d, q, z and f parameters, eqns. 6.4 assume the form

$$\left.\begin{aligned} \boldsymbol{R}_d &= \boldsymbol{R}_q = \boldsymbol{R}_a \\ \boldsymbol{L}_{dd} &= \boldsymbol{L}_{ao} + \boldsymbol{L}_{abo} + \tfrac{3}{2}\boldsymbol{L}_{a2} \\ \boldsymbol{L}_{qq} &= \boldsymbol{L}_{ao} + \boldsymbol{L}_{abo} - \tfrac{3}{2}\boldsymbol{L}_{a2} \\ \boldsymbol{L}_{df} &= \boldsymbol{L}_{afd} \\ \boldsymbol{R}_z &= \boldsymbol{R}_a \\ \boldsymbol{L}_{zz} &= \boldsymbol{L}_{ao} - 2\boldsymbol{L}_{abo} \end{aligned}\right\} \quad (6.5)$$

$\boldsymbol{L}_{afd}$ is a hybrid—see Section 3.1, page 44.

Flux linkages in the a, b and c coils can be written as follows:

$$\left.\begin{aligned} \psi_a &= \boldsymbol{L}_{aa}\boldsymbol{i}_a + \boldsymbol{L}_{ab}\boldsymbol{i}_b + \boldsymbol{L}_{ca}\boldsymbol{i}_c + \boldsymbol{L}_{af}\boldsymbol{i}_f \\ \psi_b &= \boldsymbol{L}_{bb}\boldsymbol{i}_b + \boldsymbol{L}_{bc}\boldsymbol{i}_c + \boldsymbol{L}_{ab}\boldsymbol{i}_a + \boldsymbol{L}_{bf}\boldsymbol{i}_f \\ \psi_c &= \boldsymbol{L}_{cc}\boldsymbol{i}_c + \boldsymbol{L}_{ca}\boldsymbol{i}_a + \boldsymbol{L}_{bc}\boldsymbol{i}_b + \boldsymbol{L}_{cf}\boldsymbol{i}_f \end{aligned}\right\} \quad (6.6)$$

The form is identical whether the parameters are ordinary or per-unit as long as time is normalised in the per-unit system.

The transformation of flux linkages is the same as that of voltages, and with the particular choice of the ordinary voltage transformation of eqn. 3.14, the form is identical in ordinary or per-unit parameters. In per-unit notation it is

$$\begin{bmatrix}\psi_d\\ \psi_q\\ \psi_z\end{bmatrix} = \tfrac{2}{3}\begin{bmatrix}\cos\theta & \cos(\theta-120°) & \cos(\theta-240°)\\ -\sin\theta & -\sin(\theta-120°) & -\sin(\theta-240°)\\ \tfrac{1}{2} & \tfrac{1}{2} & \tfrac{1}{2}\end{bmatrix}\begin{bmatrix}\psi_a\\ \psi_b\\ \psi_c\end{bmatrix} \quad (6.7)$$

and the d, q, z flux linkages are related to inductances and currents in the (d, q, z, f) reference frame in the following way, which is again the same in ordinary or per-unit notation. A damper winding is now shown added to each axis, for greater generality:

$$\left.\begin{aligned}\psi_d &= L_{dd}i_d + L_{df}i_f + L_{dkd}i_{kd}\\ \psi_q &= L_{qq}i_q + L_{qkq}i_{kq}\end{aligned}\right\} \quad (6.8)$$

It may be noted that the per-unit results of eqns. 6.5–6.8 are all directly comparable with results given by Concordia[10], except that there is a difference in the sign of stator currents owing to his use of a source convention. However, the intermediate steps represented by eqns 6.3 and 6.4 in ordinary parameters are a feature of the present treatment.

6.2 Ideal ratio and leakage reactance

Fig. 18 shows a simple arrangement of primary and secondary windings, each wound in two sections on an iron core with the permeability assumed infinite. The core has three narrow gaps in it with reluctances R_1, R_2, and R_3; the number of turns in the two sections of the primary winding are m_1 and n_1, and the number of turns in the two sections of the secondary winding are m_2 and n_2. This configuration may be treated as having no leakage flux in the designer's sense, since, if the gaps are made sufficiently small, any stray flux in paths external to the gaps is negligible in comparison with the flux in the gaps. All flux in the gaps links at least a part of both windings.

With 1 A flowing in the primary winding only, it follows that

$$m_1 = R_1\phi_1 + R_2(\phi_1+\phi_2) \quad (6.9)$$
$$n_1 = -R_2(\phi_1+\phi_2) - R_3\phi_2 \quad (6.10)$$

where ϕ_1 and ϕ_2 are the fluxes crossing gaps 1 and 3 in the directions shown, and returning across gap 2. The primary self inductance is calculated from the total flux linkage with the primary winding, excited by unit current; namely
$$L_{11} = m_1\phi_1 - n_1\phi_2 \qquad (6.11)$$
and the mutual inductance is calculated from the total flux linkage with the secondary winding under the same condition; namely
$$L_{12} = m_2\phi_1 - n_2\phi_2 \qquad (6.12)$$

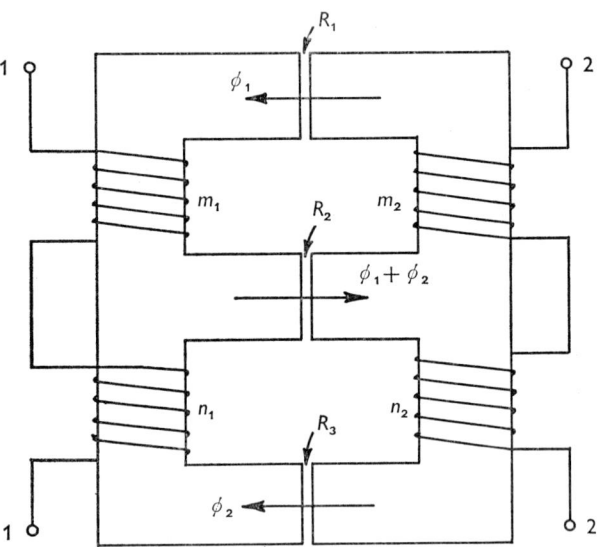

Fig. 18 Configuration of coupled coils

The turns ratio, defined by eqn. 2.39 for the particular condition of primary excitation, is
$$N_p = \frac{m_1\phi_1 - n_1\phi_2}{m_2\phi_1 - n_2\phi_2} \qquad (6.13)$$
$$= \frac{L_{11}}{L_{12}}$$
since there is no leakage flux.

Solving eqns. 6.9 and 6.10 simultaneously, and substituting into eqns. 6.11–6.13, gives the following results:
$$L_{11} = \frac{(m_1+n_1)^2 R_2 + m_1^2 R_3 + n_1^2 R_1}{R_1 R_2 + R_2 R_3 + R_3 R_1} \qquad (6.14)$$

By symmetry it follows that

$$L_{22} = \frac{(m_2+n_2)^2 R_2 + m_2^2 R_3 + n_2^2 R_1}{R_1 R_2 + R_2 R_3 + R_3 R_1} \qquad (6.15)$$

and

$$L_{12} = \frac{(m_1+n_1)(m_2+n_2)R_2 + m_1 m_2 R_3 + n_1 n_2 R_1}{R_1 R_2 + R_2 R_3 + R_3 R_1} \qquad (6.16)$$

The expression $(L_{11}L_{22} - L_{12}^2)$ can now be evaluated, and after simplification becomes

$$L_{11}L_{22} - L_{12}^2 = \frac{m_2^2 n_2^2 \left(\frac{m_1}{m_2} - \frac{n_1}{n_2}\right)^2}{R_1 R_2 + R_2 R_3 + R_3 R_1} \qquad (6.17)$$

From eqn. 6.13, the turns ratio N_p can be expressed as

$$N_p = \frac{(m_1+n_1)^2 R_2 + m_1^2 R_3 + n_1^2 R_1}{(m_1+n_1)(m_2+n_2)R_2 + m_1 m_2 R_3 + n_1 n_2 R_1} \qquad (6.18)$$

If the turns ratio is evaluated for secondary excitation (i.e. L_{12}/L_{22} since there is no leakage flux), the result is

$$N_s = \frac{(m_1+n_1)(m_2+n_2)R_2 + m_1 m_2 R_3 + n_1 n_2 R_1}{(m_2+n_2)^2 R_2 + m_2^2 R_3 + n_2^2 R_1} \qquad (6.19)$$

Some implications of the form of eqns. 6.17–6.19 are examined below.

Considering first the general forms of the equations, it is clear from eqn. 6.17 that the coupling coefficient is less than unity, although there is no 'leakage' flux in the designer's sense. The normal calculation of individual-winding leakage reactances would therefore give zero for both windings. On the other hand, the series reactances of the equivalent circuit cannot be made simultaneously zero with the coupling coefficient less than unity. This can be seen from the discussion preceding eqn. 2.45. Clearly, therefore, there can be no direct relation between leakage and series reactances in this general case.

Also, eqns. 6.18 and 6.19 give two different values for the turns ratio, and so demonstrate the point made in regard to Reference 16 in Section 2.3 and on page 33.

For the present purpose, it will be sufficient to demonstrate the conditions under which these two objections disappear, and to accept that the properties of ideal coupling, discussed in Section 2.3, then apply in total. These are:

(a) If R_1 is infinite, from eqn. 6.17 it can be seen that the coupling coefficient becomes unity. This is to be expected, since flux now only circulates around the lower half of the core forming an ideal coupling (as defined in Section 2.3) with turns ratio n_1/n_2. It can be seen from eqns. 6.18 and 6.19 that both N_p and N_s become equal to this value which is the ideal ratio. Similarly, if R_3 is infinite, the ideal turns ratio becomes m_1/m_2, and if R_2 is infinite, the ideal turns ratio becomes $(m_1+n_1)/(m_2+n_2)$.

(b) If the turns ratios are made equal on the two halves of the core; namely

$$\frac{m_1}{m_2} = \frac{n_1}{n_2} = N_i \qquad (6.20)$$

it is again clear that the coupling coefficient becomes unity, and it can be shown that $N_p = N_s = N_i$. This is an interesting case, for while eqn. 6.20 ensures that the pattern of mutual flux is unvarying, which was part of the definition of ideal coupling and ratio in Section 2.3, not all turns of the primary and secondary windings are perfectly linked by all tubes of mutual flux.

This demonstrates that the definition of ideal coupling adopted in Section 2.3 is sufficient, but is capable of extension to include slightly less restricted forms of coupling without loss of the required properties. Discussion of this matter can readily be pursued, but for the present purpose is not necessary. In the range of practical devices of interest, couplings which are ideal conform to the restricted definition.

6.3 Equivalent circuits of synchronous machines with nested damper windings

Fig. 19 shows the general, transient equivalent circuits for the direct and quadrature axes of a synchronous motor, which has an assumed full-pitched field winding on the direct axis, and three nested damper circuits —none of which are full-pitched—on each axis. The nomenclature here and in Fig. 20 is only correct for transformation eqns.3.1 and 3.2 and for the 'full-pitch X_{ad} base' defined in Section 3.2.2.

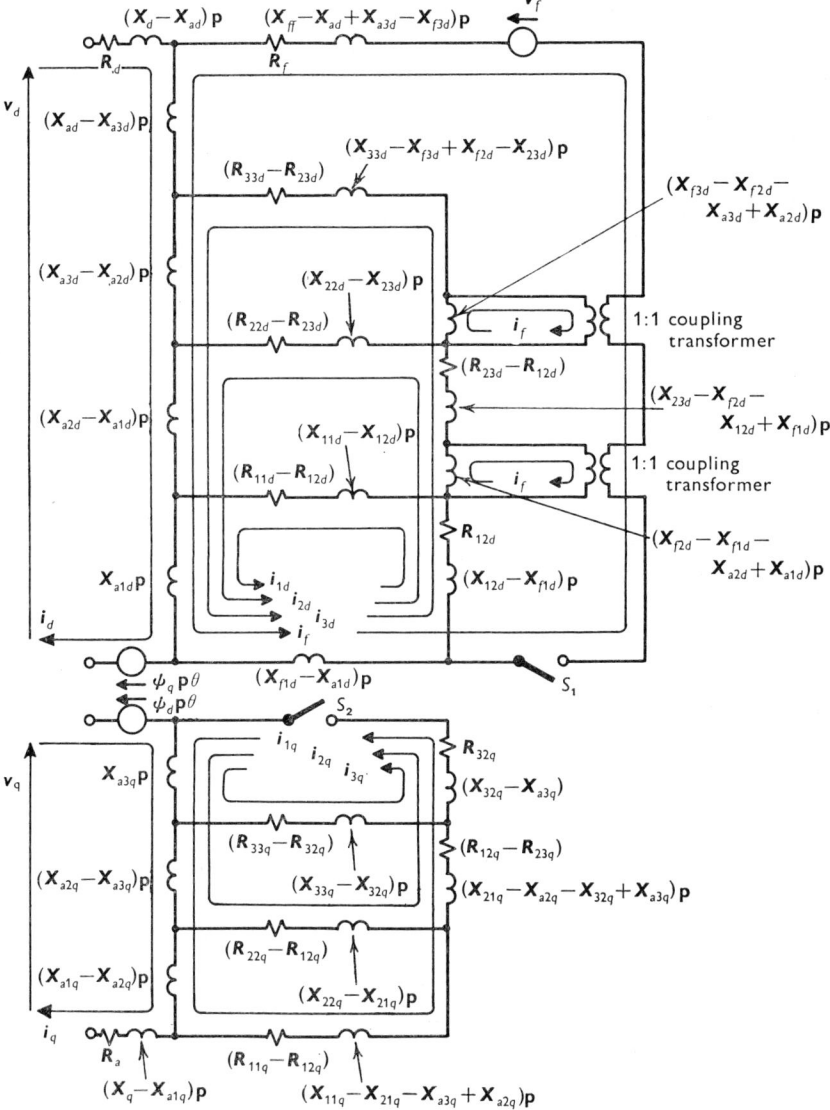

Fig. 19 Transient equivalent circuits for direct and quadrature axes of synchronous motor

Full-pitch field winding on direct axis, and three nested damper circuits on each axis are assumed. The transformation equations (eqns. 3.1 and 3.2) and the 'full-pitch X_{ad} base' have been used

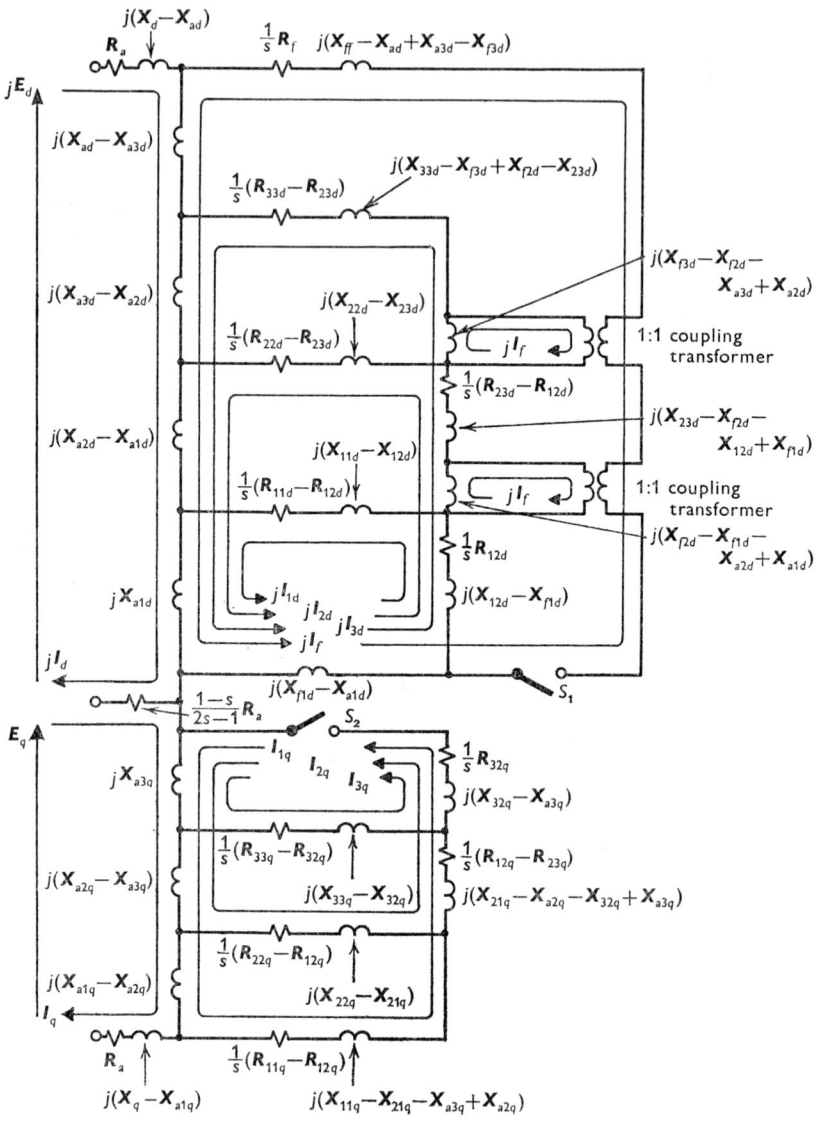

Fig. 20 Circuits of Fig. 19 simplified to describe asynchronous operation at constant slip s

The numbering of the damper circuits corresponds to that of the bars of the squirrel cage of Fig. 11; i.e. the current of the innermost circuit on the d axis flowing in bars 1 is i_{1d}, and that of the outermost included here is i_{3d}. On the q axis, bars 3 form the innermost circuit, and hence its current is i_{3q} while that of the outermost q circuit is i_{1q}. This numbering convention is the same as that used by Lewis.[16] Rankin uses a different system, numbering each bar twice—once for its effect on the d axis and once for the q axis.

The circuits appear at first sight very complicated, but when considered mesh by mesh they are readily seen to correspond to the familiar 2-axis equations of a field and nested circuits coupled via the sine wave of air-gap flux to the stator. For example, for the left-hand mesh of the d axis

$$v_d = (R_a + X_d \mathrm{p}) i_d + X_{ad} \mathrm{p} i_f + X_{a3d} \mathrm{p} i_{3d} + X_{a2d} \mathrm{p} i_{2d} + X_{a1d} \mathrm{p} i_{1d} - \psi_q \mathrm{p}\theta$$

and for the second damper circuit of the d axis

$$0 = (R_{22d} + X_{22d} \mathrm{p}) i_{2d} + (R_{23d} + X_{23d} \mathrm{p}) i_{3d} + (R_{12d} + X_{12d} \mathrm{p}) i_{1d}$$
$$+ X_{f2d} \mathrm{p} i_f + X_{a2d} \mathrm{p} i_d$$

The 1:1 ideal coupling transformers are necessary when two or more damper circuits are represented on the direct axis. This is because the field winding is inductively coupled to all the d-axis circuits, but is not conductively coupled to them. An ideal voltage generator v_f is included in the circuit to represent the d.c. excitation of the field circuit. Since the coupling transformers are ideal, the establishment of current i_f in one winding inevitably sets up an equal current in the other, and this equality is retained at all times; also a voltage across one side is always reflected in an equal one on the other side. The analysis of the field-circuit mesh is therefore as follows:

$$v_f = R_f i_f + (X_{ff} - X_{ad} + X_{a3d} - X_{f3d}) \mathrm{p} i_f + (X_{f3d} - X_{f2d} - X_{a3d}$$
$$+ X_{a2d}) \mathrm{p}(i_{3d} + i_f) + (X_{f2d} - X_{f1d} - X_{a2d} + X_{a1d}) \mathrm{p}(i_{2d} + i_{3d} + i_f)$$
$$+ (X_{f1d} - X_{a1d}) \mathrm{p}(i_{1d} + i_{2d} + i_{3d} + i_f) + X_{ad} \mathrm{p}(i_d + i_f)$$
$$+ X_{a3d} \mathrm{p} i_{3d} + X_{a2d} \mathrm{p} i_{2d} + X_{a1d} \mathrm{p} i_{1d}$$

therefore

$$v_f = (R_f + X_{ff} \mathrm{p}) i_f + X_{f3d} \mathrm{p} i_{3d} + X_{f2d} \mathrm{p} i_{2d} + X_{f1d} \mathrm{p} i_{1d}$$

Any number of nested damper windings which are coupled conductively through their mutual resistances in the end rings can be represented without coupling transformers.

The switch S_2 is included in the q-axis circuit at the point equivalent to the centre of the cage about that axis; i.e. in series with the common mutual resistance $\boldsymbol{R}_{32q}$. The opening of this switch corresponds to no interconnection of the damper winding between the poles; it does not correspond to the absence of an effective damper winding on the q axis, but simply imposes the condition

$$\boldsymbol{i}_{1q}+\boldsymbol{i}_{2q}+\boldsymbol{i}_{3q} = 0$$

For the particular case of only one damper winding on each axis, the circuits reduce to those of Fig. 10.

The two circuits described above are shown for the general transient condition and are not independent, because of coupling through the motional voltages $\boldsymbol{\psi}_q \boldsymbol{p}\theta$ and $\boldsymbol{\psi}_d \boldsymbol{p}\theta$. This coupling has been shown in an interesting and useful way by Lewis[16] for the particular conditions of constant slip and zero-field voltage, and the resulting circuit for the preferred per-unit system and transformation, which unlike Lewis's system has the q axis leading the d axis, is shown in Fig. 20. This circuit is useful in investigating the asynchronous performance of synchronous motors, and is discussed fully by Lewis. It is derived by representing the axis voltages as slip-frequency phasors, and in the form used here they are

$$\boldsymbol{V}_d = \hat{\boldsymbol{v}}_a$$
$$\boldsymbol{V}_q = j\hat{\boldsymbol{v}}_a$$

The most interesting feature of the circuit is the elimination of the motional-voltage generators in Fig. 19 by the coupling resistance $(1-s)\boldsymbol{R}_a/(2s-1)$, and the division of all rotor resistances by the fractional slip s. This coupling resistor is, of course, negative for $s < 0.5$. It is also interesting to note, as Lewis does, that in all machines for $s = 0.5$, (when this resistance is infinite), and for balanced induction motors, (in which the d and q axes are identical), no current flows in this branch, since $\boldsymbol{I}_d = j\boldsymbol{I}_q$.

7 REFERENCES

1 SHEPPARD, D. G.: 'Elements of fluid mechanics' (Harcourt, Brace & World, 1965) Chap. 5
2 RANKIN, A. W.: 'Per-unit impedances of synchronous machines', *Trans. Amer. Inst. Elect. Engrs.*, 1945, **64**, pp. 569–572
3 RANKIN, A. W.: 'Per-unit impedances of synchronous machines—Pt. 2', *ibid.*, 1945, **64**, pp. 839–841
4 PARKHURST, R. C.: 'Dimensional analysis and scale factors' (Chapman & Hall, 1964)
5 STEPHEN, D. D.: 'Co-ordinated per-unit systems for electrical machine characteristics', *Elect. Energy*, 1957, **1**, pp. 438–445
6 GIBBS, W. J.: 'Electric machine analysis using matrices' (Pitman, 1962), p. 6
7 WHITE, D. C., and WOODSON, H. H.: 'Electromechanical energy conversion' (Wiley, 1959), p. 522 *et seq.*
8 ADKINS, B.: 'The general theory of electrical machines' (Chapman & Hall, 1957), pp. 6, 19, 116
9 KRON, G.: 'Tensors for circuits' (Dover, 1959)
10 CONCORDIA, C.: 'Synchronous machines' (Wiley, 1951), p. 20
11 ROTHE, F. S.: 'Introduction to power systems analysis' (Wiley, 1953), p. 76
12 FITZGERALD, A. E., and KINGSLEY, C.: 'Electric machinery' (McGraw-Hill, 1961, 2nd edn.)
13 SEELY, S.: 'Electromechanical energy conversion' (McGraw-Hill, 1962)
14 KIMBARK, E. W.: 'Power system stability—Vol. 1' (Wiley, 1948), p. 54
15 CRARY, S. B.: 'Power system stability—Vol. 1' (Wiley, 1945), p. 278
16 LEWIS, W. A.: 'A basic analysis of synchronous machines—Pt. 1', *Trans. Amer. Inst. Elect. Engrs.*, 1958, **77**, pp. 436–456
17 KRON, G.: 'Tensor analysis of networks' (Macdonald, 1965), pp. 462–70
18 Proposed standard, 'Definitions of basic per-unit quantities for alternating-current rotating machines', Amer. Inst. Elect. Engrs., proposed standard 86, Feb. 1961
19 BILLIG, E.: 'The calculation of the magnetic field of rectangular conductors in a closed slot, and its application to the reactance of transformer windings', *Proc. IEE*, 1951, **98**, (Pt. 4), pp. 55–64

20 CONCORDIA, C.: 'Relations among transformations used in electrical engineering problems', *Gen. Elect. Rev.*, 1938, pp. 323-325
21 DOHERTY, R. E., and NICKLE, C. A.: 'Synchronous machines—Pt. 4', *Trans. Amer. Inst. Elect. Engrs.*, 1928, **47**, pp. 457-492
22 DOHERTY, R. E., and NICKLE, C. A.: 'Three-phase short-circuit synchronous machines,' *ibid.*, 1930, **49**, pp. 700-714
23 WIESEMAN, R. W.: 'Graphical determination of magnetic fields', *ibid.*, 1927, **46**, pp. 141-148
24 RANKIN, A. W.: 'The direct and quadrature axis equivalent circuits of the synchronous machine,' *ibid.*, 1945, **64**, pp. 861-867
25 KILGORE, L. A., 'Calculation of synchronous machine constants', *ibid.*, 1931, **50**, pp. 1201-1213
26 MATHUR, R. M.: 'Starting performance of segmental rotor reluctance machines'. Ph.D. thesis, University of Leeds, 1969
27 DOHERTY, R. E., and NICKLE, C. A.: 'Synchronous machines—Pt. 1', *Trans. Amer. Inst. Elect. Engrs.*, 1926, **45**, pp. 912-942
28 PARK, R. H.: 'Two reaction theory of synchronous machines', *ibid.*, 1929, **48**, pp. 716-727
29 LINVILLE, T. M.: 'Starting performance of salient-pole synchronous motors', *ibid.*, 1930, **49**, pp. 531-547
30 ALGER, P. L.: 'The nature of polyphase induction machines' (Wiley, 1957), p. 188
31 LANGSDORF, A. S.: 'The theory of a.c. machinery' (McGraw-Hill, 1955), pp. 363-387
32 VICKERS, H.: 'The induction motor' (Pitman, 1949), pp. 105-133
33 JONES, C. V.: 'The unified theory of electrical machines' (Butterworth, 1967)
34 MAJMUDAR, H.: 'Electromechanical energy convertors' (Allyn & Bacon, 1965), pp. 454-460
35 ROGERS, G. J.: 'Linearised analysis of induction-motor transients', *Proc. IEE*, 1965, **112**, (10), pp. 1917-1926
36 LAWRENSON, P. J., and STEPHENSON, J. M.: 'Note on induction-machine performance with a variable-frequency supply', *ibid.*, 1966, **113**, (10), pp. 1617-1623
37 REED, M. B., and ROBERGE, R. M.: 'Generalisation of the normalisation (per-unit) technique', *IEEE Trans.*, 1969, **PAS-88**, pp. 1665-1672.

INDEX

a.c. commutator machines, 87, 88, 102
Adkins, 12, 13, 36, 38, 39, 66, 68, 90, 94, 97, 98
advantages of per-unit system, 7, 8
air-gap flux distribution, 10, 13, 48, 54, 63, 68, 77, 83, 87, 100, 101
Alger, 77
associated per-unit system, 15, 92–94

base quantities
  general, 4, 7
  current, 10, 16, 19, 20, 24, 42, 57, 79, 81, 92
  dependent, 4, 17, 19, 22, 95, 96, 97
  flux linkage, 95, 96
  frequency, 95
  impedance, 16, 19, 20, 24
  inertia, 98, 100
  power and apparent power, 14, 15, 17, 18, 19, 21, 81, 91, 94, 97, 100, 101
  reactance, 16, 20, 95
  standard definitions, 14, 91
  time, 44, 94, 100, 105
  torque, 97, 98
  speed, 94, 100
  voltage, 16, 19, 24, 42, 57, 79, 92
  voltage–current product, 18, 23, 42–53, 57, 62, 79, 86, 91, 101
base systems
  equal mutuals, 11, 12, 73
  full-pitch X_{ad}, 71, 100, 101, 108
  m.m.f., 10, 12, 72
  unit voltage, 10, 74
  X_{ad}, 10, 12, 13, 63–72, 77, 83, 88, 102
Billig, 32

Concordia, 12, 13, 36, 38, 43, 44, 46, 66, 68, 90, 94, 98, 105
coupling coefficient, 35, 107
Crary, 4
current base, 10, 16, 19, 20, 24, 42, 57, 79, 81, 92

damper circuits, 68–72, 73, 74, 100, 105 108–112
d. c. machine, 89, 102
dimensions, dimensionless parameters, 3, 6
Doherty, 43, 75, 94

equal-effect principle, 19, 48, 49, 54, 63, 66, 75, 76, 79, 80, 87, 101, 102
equal-mutuals base, 11, 12, 73
equivalent circuits, 21, 28, 50, 51, 52, 68, 69, 74, 80, 85, 100, 108–112

Fitzgerald, 14, 43
flux linkage, 50–53, 95, 96
flux density, distribution fundamental to machine operation, 10, 13, 48, 54, 63, 68, 77, 83, 87, 100, 101
  space harmonics of flux density, 64
frequency base, 95
full-pitch X_{ad} base, 71, 100, 101, 108

Gibbs, 11, 23, 44, 84, 90

harmonics, 64

ideal field pattern, 29
ideal turns ratio, 29–35, 64–68, 70–72, 101, 105–108
impedance base, 16, 19, 20, 24
inductances, negative, 34, 68, 72, 73, 75
inductance base, 16, 19, 20, 24, 67, 95
induction machine
  polyphase, 9, 77–81, 92, 99, 102
  1-phase, 81–87, 102
inertia constant, 98, 100

Jones, 84, 85

Kilgore, 68
Kimbark, 4

Kingsley, 14, 43
Kron, 14, 23

Langsdorf, 84
Lawrenson, 78
leakage inductance, 8, 27–35, 49, 64, 67, 80, 85, 101, 105–108
Lewis, 13, 15, 20, 27, 32, 34, 43, 46, 47, 50, 90, 94, 107, 111, 112
Linville, 73, 94, 95

Majmudar, 86
Mathur, 70
m.m.f. base, 10, 12, 72
m.m.f. wave, flat-topped, 10, 72
moment of inertia, 98, 100
mutual inductance
  asymmetry of, 17, 18, 23, 38
  normalisation of, 16, 20

Nickle, 43, 75, 94
normalisation, 3, 14, 24, 26, 39, 42, 101

'ordinary' parameters, 6, 15

Park, 75, 94, 95
Parkhurst, 3, 7
peak base quantities, 10, 13, 14, 15, 42, 57, 90, 93, 100
permeability
  infinite, 70, 105
  finite, 30, 70
physical model, 4, 26, 39, 41, 44, 53, 55, 56, 58–63, 101
power base, 14, 15, 17, 18, 19, 21, 81, 91, 94, 97, 100, 101
power-invariant transformations, 23, 25, 26, 39, 55, 58, 84
power systems, 4, 13, 59, 62, 92

Rankin, 6, 10, 12, 13, 15, 19, 44, 46, 66, 67, 68, 72, 75, 90, 94
rated values, 12, 25
reactance base, 16, 20, 95

Reed, 5, 19, 27
r.m.s. base quantities, 14, 15, 18, 21, 90, 93
Roberge, 5, 19, 27
Rogers, 73, 79
Rothe, 4, 13, 14, 15, 27, 33, 36, 58, 59, 90, 93, 94, 98

saturation, 64, 70
Seely, 27
series reactances, 8, 27–35, 49, 64, 67, 80, 85, 101, 105–108
Sheppard, 3, 7
Speed base, 94, 100
standard definitions, 14, 91
Stephen, 9, 10, 14, 75, 81, 95
Stephenson, 78
symbols, list of, 1–3
symmetrical components, 58–63

time base, 44, 94, 100, 105
torque base, 97, 98
transformations
  orthogonal, 46, 47, 55, 58, 86, 103
  power-invariant, 23, 25, 26, 39, 55, 58, 84
  symmetrical component 58–63
  2-phase/2-axis, 55–58, 82–86
  3-phase/2-axis, 8, 13, 25, 36–53, 78, 94, 99, 103
transformer, 8, 14, 16, 27
turns ratio, 10, 18, 19, 20, 26, 27–35, 43–53, 101

Vickers, 85
voltage base, 16, 19, 24, 42, 57, 79, 92

White, 12, 74, 90
Wieseman, 66
Woodson, 12, 74, 90

X_{ad} base, 10, 12, 13, 63–72, 77, 83, 88, 102

TK
2189
H37

NOV 19 1971